Management for Professionals

More information about this series at http://www.springer.com/series/10101

JinHyo Joseph Yun

Business Model Design Compass

Open Innovation Funnel to Schumpeterian New Combination Business Model Developing Circle

Springer

JinHyo Joseph Yun
Tenured Researcher of DGIST
and Professor of Open Innovation
Academy of SOItmC
Daegu, Korea, Republic of (South Korea)

ISSN 2192-8096 ISSN 2192-810X (electronic)
Management for Professionals
ISBN 978-981-10-4126-6 ISBN 978-981-10-4128-0 (eBook)
DOI 10.1007/978-981-10-4128-0

Library of Congress Control Number: 2017941317

Printed on acid-free paper

This Springer imprint is published by Springer Nature
The registered company is Springer Nature Singapore Pte Ltd.
The registered company address is: 152 Beach Road, #21-01/04 Gateway East, Singapore 189721, Singapore

Preface

How do we conquer the growth limits of a global capitalist economy?

I have studied open innovation and open business model to answer this question nearly for 10 years. The creative three books such as *Open Innovation* (2003), *Open Business Models* (2006), and *Open Services Innovation* (2011) from Henry Chesbrough who is the founder of open innovation gave me big implications.

I realized the requirement of a global research community and journal to answer my research question in 2010. So I prepared a research community and a research journal to answer my only one research question for 5 years with more than 1000 professors and researchers from 35 countries.

At first, I made the *Journal of Open Innovation: Technology, Market, and Complexity* (JOItmC) with nearly 100 professors and researchers from 25 countries in 2014. Please visit www.jopeninnovation.com to see several great papers from this open access journal. And the Society of Open Innovation: Technology, Market, and Complexity (SOItmC) was established in 2015 by 500 professors and researchers from 30 countries. I hosted the SOItmC and KCWS 2015 conference at DGIST on June 14–18, 2015, with the theme "Open Innovation, Knowledge City and Creative Economy." The SOItmC and CSCOM 2016 conference was hosted by San Jose State University on May 31–June 3, 2016, with 127 great papers under the theme "Open Innovation of Start-Ups and Firms in Value Chain." Please visit www.openinnovationtmc.org to see details of conferences.

I prepared this book to answer to the requirements for a guidance to connect logically and practically between open innovation and business model from several professors, researchers, and firms' CEOs who joined at SOItmC and JOItmC.

First, I thank a lot the ex-president of DGIST, Mr. President, Dr. SungChul Shin. He gave chances for me to research deeply on open innovation and business model. Second, I thank a lot honor professor Henry Chesbrough. He gave me several teachings directly and indirectly for me to develop my own research frameworks for open innovation and open business model. Third, I thank a lot all expanded editor board members of JOItmC and all members of SOItmC including honor professor Philip Cooke from Bergen University College, Norway; honor professor Fumio Kodama, University of Tokyo, Japan; honor professor Fred Phillips, Stony Brook University, USA; honor professor Venni Krishna, Jawaharlal Nehru University, India; honor professor Francisco Javier Carrillo Gamboa, Tecnológico de Monterrey, Mexico; honor professor Anil K. Gupta, Indian Institute of

Management, India; honor professor Keld Laursen, Copenhagen Business School, Denmark; honor professor Keun Lee, Seoul National University, Korea; honor professor KongRae Lee, DGIST, Korea; honor professor Loet Leydesdorff, University of Amsterdam, Netherlands; and honor professor Ulrich Witt, Max Planck Institute of Economics, Germany.

February 27, 2017 JinHyo Joseph Yun

Acknowledgment

This book was supported by the DGIST R&D Program of the Ministry of Science, ICT and Future Planning (16-IT).

Contents

Recommendation

1. **Professor Loet Leydesdorff**

Professor, University of Amsterdam
Amsterdam School of Communication Research (ASCoR)

> JinHyo Joseph Yun extends the concept of "open innovation" to the complex dynamics of technology, markets, and control. This Triple Helix is worth investigating: how are new options for innovation generated?

2. **Professor Venni Krishna**

Professorial Fellow, University of New South Wales, Sydney, and Editor-in-Chief, *Science, Technology and Society* (Sage Publications)

> In the footprints of Henry Chesbrough on Open Innovation, Professor JinHyo Joseph Yun has made relentless intellectual efforts to advance our knowledge on this theme for the last five years. This new volume on *Business Model Design Compass* is indeed an important step in this direction.

3. **Professor Yigltcanlar Tan**

Associate Professor, Queensland University of Technology, and Editor, *Sustainability*

> This book tackles the ever growing challenge of successful business model design in the age of global knowledge economy rivalry. It generates new expansions for our understanding on how open innovation should be factored in the business model design process.

4. **Professor KongRae Lee**

Professor at graduate school, and Director of Management of Innovation Program of DGIST. Honorary founder and editor of *Asian Journal of Technology Innovation*

> This book introduces the way from open innovation to creative business model which is the key requirement in the 4th industrial revolution era. The convergence innovation which goes with open innovation will motivate emergent new business model.

5. **Professor Anil K. Gupta**

Professor, Indian Institute of Management, Ahmedabad, the founder of the Honey Bee Network

> With increasing disruption in global business environment due to arrival of innovation hungry start-ups and other actors, need for companies to recalibrate their innovation compass was never so high. This book makes a very useful contribution to the future discourse on open innovation in existing organizations by harnessing entrepreneurial potential of all members of supply chains. Linking intra-preneurship with supporting innovations based enterprises outside is a novel contribution. Policy makers, industrial and technology park leaders, and business leaders wanting to revitalize the learning environment will find this book very useful.

6. **Professor Ulrich Witt**

Professor at Griffith Business School of Griffith University (Australia), Director of Evolutionary Economy Group at the Max Planck (Germany)

> Open innovation processes have become an essential driver of the ICT revolution, particularly in the software industry. In his inspiring book, Dr. Yun develops fruitful concepts and strategies for expanding the success story of open innovations into new applications and the corresponding business models.

7. **Fred Philips**

Professor at Department of Technology and Society of Yuan Ze University (Taiwan), Editor in Chief of "Technological Forecasting and Social Change"

> IT has enabled faster and faster recombination of links in value chains, creating new chains and enabling new business models, including open innovation models. JinHyo Joseph Yun stands out among the creative thinkers envisioning the future of capitalism, by creating upbeat, plausible scenarios of future economic systems utilizing open innovation business models.

Part I

Introduction

Introduction

1

Abstract

Here, I begin by discussing the requirements of open innovation, after which I start to develop our idea from the open innovation to the business model developing circle. Next, I discuss how to vitalize open innovation conceptually. Diverse strategies to motivate open innovation will be discussed following this. Third, we look into the locus of open innovation. I will find ways to combine the technology and the market from the locus of open innovation, that is, to say, forge a creative business model. Lastly, I will explain how to measure open innovation. I can understand the reality of open innovation and the business model from concrete measures.

Keywords

Locus of open innovation • Measure of open innovation

This chapter is directly based on the following three papers:

Yun J.H.J (2014), Why Do We Need Open Innovation? *Asia Pacific Journal of Innovation and Entrepreneurship. Vol. 8, No. 1*

Yang and Yun (Corresponding author) (2014), New Perspectives on Open Innovation from Game Theory: The Prisoners' Dilemma and the Ultimatum Game. *Asia Pacific Journal of Innovation and Entrepreneurship. Vol. 8, No. 1*

Yun, Avvari, Jeong, and Lim (2014), Introduction of an objective model to measure open innovation and its application to the information technology convergence sector. *International Journal of Technology, Policy and Management. Vol. 14, No. 4*

J.J. Yun, *Business Model Design Compass*, Management for Professionals,
DOI 10.1007/978-981-10-4128-0_1

1.1 The Requirements and Method to Vitalize Open Innovation

As the knowledge-based economy develops, the amount of knowledge that is circulated rapidly increases. Therefore, it is becoming necessary for firms to increase their use of external knowledge and technologies (Yun and Mohan 2012; Yun et al. 2013) and for governments to arrange policies to bolster open innovation strategies among firms (Yun 2010; Yun and Park 2012). The literature on strategic alliances, virtual corporations, buyer–supplier collaborative relationships, and technology collaborations continuously increases, which indicates the importance of external integration and outsourcing (Teece et al. 1997).

Earlier, Henry W. Chesbrough tried to explain such patterns of corporate activities with the colossal concept of open innovation (Kim 2009). Summing up these issues, aside from large companies such as IBM, 3M, and Intel (Chesbrough 2003), small companies should consider open innovation strategies as essential to their survival rather than as merely an option (Laursen and Salter 2006; Vad de Vrande et al. 2009; Yun and Jung 2013; Yun and Mohan 2012; Yun and Park 2012). As the knowledge-based society progresses, the necessity of open innovation strategies continues to grow along with such issues as technology transfers and industry–university collaborations.

1.1.1 Why Do We Need Open Innovation?

First, as the knowledge-based economy develops, the areas of knowledge such as protected knowledge, protectable knowledge, and normal knowledge have all increased in the economic system. Therefore, any firm can easily and inexpensively obtain technologies that are required for the firm to innovate with modern products and processes or to introduce new business models (Foray and Lundvall 1998).

Second, the growth energy of the globally weakened economy can be recovered through open innovation. Surplus value or enterprise returns are diminishing. According to Kondratieff, despite the economic fluctuations, every advanced capitalized economy system in the end arrives at the top of the growth chart. The concrete evidence of this is that basic interest rates, which indicate the price of capital, are approaching an actual value of zero (Freeman 2011). The world's economically developed major countries are showing true zero as the prevailing interest rate. In this zero-interest-rate era, open innovation can be a new trigger for economic growth. Open innovation lets firms find new breakthroughs by allowing them to look for new markets outside of their firm for their technology or for new technologies outside of the firm for new or more modern markets. Korea's creative economic policy has several sub-policies that are based on open innovation.

Third, the power law in the economy can be conquered by open innovation (Laherrere and Sornette 1998). Currently, a small number of large companies obtain the majority of economic benefits in most growing and mature industries. Open innovation can give new entrants the chance to compete continuously with these companies in new and growing industries. In the end, open innovation will make new firms to continuously emerge in new sectors and in the economy. The long tail phenomena can occur in diverse industries through the open innovation paradigm (Elberse 2008).

Fourth, we can construct a creative economy through open innovation. A creative economy requires creative new products and processes introduced by new firms in various emerging sectors. Open innovation can allow a creative connection between the technology and the market. As a result, several creative new products and processes will appear in the market. In the end, creative firms can appear, grow, and construct the creative economy (Howkins 2002; Markusen et al. 2008).

1.1.2 How Can We Vitalize Open Innovation at the National Level or Higher?

First, we should conquer two cultures, referring to the differentiation between the humanities or social science culture and the engineering and natural science culture (Snow 1959). Mutual penetration between the two cultures can yield creative results in several industries, engineering sectors, and social sciences. If we conquer this two-culture duality, as the Western countries and Japan started to do in the 1950s–1960s, we can vitalize open innovation through a convergence or fusion between the humanities or social sciences and engineering or the natural sciences (Stokes 1997; Tushman and O'Reilly 2007).

Second, patent trolls should be allowed and should be constructed by universities, national labs, and firms (McDonough III 2006; Risch 2012). Patent trolling means that new technologies that are developed at universities and national labs can be compulsorily transferred to firms. In addition, every researcher in every national lab and every professor at every university will have a few more chances to commercialize their technologies through patent trolling. If intellectual ventures have a few more chances to buy patents from national labs and universities, which receive research funds from the government, the commercialization of new technology will increase dramatically. We can find evidence of this in the case of the USA.

Third, business model's patent production and utilization should be encouraged. Business model patents are creative bridges between the technology and the market. Accordingly, an increase in the number of business model patents will mean that more technologies can meet new markets creatively (Calia et al. 2007; Chesbrough 2012; Zott and Amit 2008). According to increasing trends in US business model patents, business models concretely connect technologies and markets. Examples include the one-click patent or cloud-sourcing patent of Amazon or the escrow patent of eBay. These are examples by which any business model patent can create a specific industry and forge a new large business (May and Flint 2009; Harinarayan et al. 2010; Hartman et al. 1999). Autonomous cars or the intelligent robot industry is appearing based on diverse business model patents (Yun et al. 2016).

1.2 Locus of Open Innovation

However, the fact that Arrow's information paradox (Arrow 1962; Dosi et al. 2006), referring to the discordance existing between technology demanders and suppliers; chasm (Levinthal and Rerup 2006; Moore 2002; Shove 1998; Sroufe et al. 2000), referring to the incomplete transfer of technologies to markets; and "death valley" (Auerswald and Branscomb 2003; Moran 2007; Rai et al. 2008), referring to large gaps between technologies and markets, are not overcome proves that many technologies are still ineffectively connected to markets.

As a result, it is necessary to understand the reasons behind the inactive cooperation between businesses and the difficulties in overcoming Arrow's information paradox in the real world despite those movements for the activation of open innovations that have been aggressively made. The prisoner's dilemma game directly shows that in situations in which defection is rewarded more, players cannot go toward Pareto optimality.

Figure 1.1 shows the payoff matrix of firms A and B according to their open and closed innovation strategies. The value q is defined to indicate the size of common knowledge when both firms A and B choose open innovation strategies. It is set to 0.5. In addition, the value r implies the value of the advantage of a technical monopoly that will be lost by a firm if it selects open innovation strategies when the competing firm selects closed innovation strategies (Porter 1985); it is set to 0.2. Moreover, as additional variables to note, α_{ij} and βij are considered, where α_{ij} is the cost to a firm, j, that is required for combining firm i's knowledge with that of j; and βij is a variable describing the synergetic benefits to firm j arising from the combination of firm i's knowledge with its own knowledge. It is assumed that the payoffs A and B, which each firm earns by participating in the game, are set to 1 equally. This logic of the payoff matrix originates from the idea that the "loss of monopolistic profits" is larger than the loss brought about by "commodity traps" in the short run. Therefore, although the cases where both firms select "Open" are the most desirable in terms of the payoffs of both firms, individual firms select "Closed," which is a closed innovation strategy, as their best strategy. Eventually, the result converges on (Closed, Closed) for firms A and B, which is a Nash equilibrium below Pareto optimality. Hence, with this structure, the prisoner's dilemma game is quite proper as a research frame (Axelrod and Hamilton 1981) and embodies well the problems of achieving mutual cooperation (Rapoport 1965) herein (Open, Open). As such, selecting "Closed" is always better than selecting "Open," at least in the one-shot prisoner's dilemma, which is why companies today face Arrow's information paradox.

However, in business environments, companies in identical or different industries are often connected to each other for their business and are therefore very likely to meet repeatedly, assuming that they do not go bankrupt. It is thus the iterated prisoner's dilemma that is to be applied in this case as the research methodology. Moreover, it is impossible to guess in advance how many interactions with the

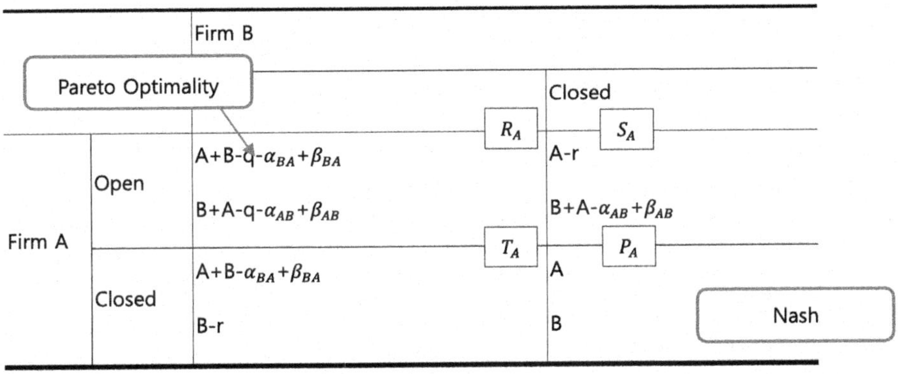

Fig. 1.1 A short-term prisoner's dilemma game of two firms (Source: Yang and Yun (2014))
q = the amount lost by common knowledge sharing between the two firms
r = the value loss of the advantage of a technical monopoly
α_{ij} = a cost to firm j that is required for combining i's knowledge with its own
βij = a synergy benefit to firm j arising from the combination of i's knowledge with its own

same business partner would continue; hence, unlike in the one-shot game, defections can no longer be the only stable solution, especially when the probability that companies meet again in the future is high (Axelrod 1986). In addition, with the probability at a high level, cooperation based on reciprocity can thrive even in the grim noncooperative world (Axelrod and Hamilton 1981).

There are many studies that support this study's basic assumption that if firms use open innovation strategies based on the repetitiveness and reciprocity of interactions between firms as such, they can escape the Nash equilibrium based on short-term perspectives and move toward Pareto optimality. Robert M. Axelrod and William D. Hamilton, who conducted studies regarding cooperation over the long term, noted repetitiveness and reciprocity as indispensable requisites for mutual cooperation to be continuously maintained (Axelrod and Hamilton 1981). Meanwhile, reciprocity is divided into direct reciprocity (Axelrod and Hamilton 1981; Nowak et al. 1995; Trivers 1971) and indirect reciprocity. In particular, if agents have opportunities to gain reputations in situations of the latter, cooperation can be developed effectively, even in relationships among multiple agents, unlike the results of previous studies (Alexander 1987; Cave 1984; Joshi 1987; Leimar and Hammerstein 2001; Lotem et al. 1999; Nowak and Sigmund 1998; Taylor 1976; Wedekind and Milinski 2000). That is, the possibility of the successful formation of cooperation between firms or of firms' open innovation in today's situation, in which numerous firms exist in the world of business, means that even those firms that have been caught in a trap of short-term interactions, such as Arrow's information paradox, can begin to accomplish Pareto optimality.

1.3 Concrete Existence of Open Innovation

1.3.1 Diverse Measures of Open Innovation

When the concept of open innovation was proposed by Chesbrough (2003), it was not conceptualized in a measurable form but was based on various forms of case studies. He also considered firms with trade patents as open innovation firms and conducted in-depth analyses of these firms. During the process of analyzing various open innovation business models, Chesbrough et al. (2006) specifically extended the concept of the open innovation of firms to a business model concept beyond open innovation strategies.

Laursen and Salter (2006) were the first to suggest the concept of the comparable measurement of open innovation of firms beyond the level of case analysis. For the measurement of the open innovation levels of firms, they developed the concepts of the "width" and "depth" in open innovation and attempted to measure them for the first time. Laursen and Salter (2006) analyzed the relationship between open innovation and firm performance regarding manufacturing enterprises in the UK. Their analysis was based on data collected through a technical innovation survey of what were considered advanced enterprises, and it was based on concepts from the Oslo Manual. They measured the width of open innovation based on the answers of firms that had open innovation levels of more than two points on a five-point scale. They measured the "depth of open innovation" based on the answers from firms that had open innovation levels of more than three on a five-point scale. Although these measures developed by Laursen and Salter (2006) provide a foundation for the measurement of the open innovation levels of firms, they have limitations. Laursen and Salter measured open innovation levels based on a survey in which the responses were obtained from the firms themselves; thus, the results were not based on objective data. The correlation between the width and depth of open innovation showed a very high level of 0.417; thus, there may be the mixed use of the different concepts of open innovation width and open innovation depth.

As discussed above, many studies of open innovation have been conducted using case study methods and interviews, while others have adopted survey-based research methods following the framework introduced by Laursen and Salter (2006), as shown in Table 1.1. In other papers, financial data have been utilized to measure firm performance (Lee et al. 2016; Yun et al. 2009). However, because objective open innovation measurement instruments have not been sufficiently developed or suggested, quantitative data-based research, such as comparisons between open innovation levels among firms and at the nation level, remains insufficient.

Yun and Mohan (2012) slightly improved the open innovation measuring method of Laursen and Salter (2006). First, they conceptualized open innovation levels on a single measurement basis by multiplying the open innovation's "width" and "depth." In addition, they improved the concept of the open innovation depth so that it would be more suitable for its logical definition by arithmetically calculating the mean value of the answers on a scale of five obtained from firms in surveys on open innovation levels. However, this approach does not overcome the fundamental problem

Table 1.1 Measure of open innovation in research

Literature review of OI	Survey	Interview
Chesbrough (2003, 2006)		AT&T, Bell Labs IBM, etc.
Laursen and Salter (2006)	Width of OI; depth of OI	
West and Gallagher (2006)		Interview Qualitative data
Vrande et al. (2009)	EIM; survey data	
Lichtenthaler (2009)	Survey; financial DB	
Hughes and Wareham (2010)		Interview, documents, observation
Chiaroni et al. (2010)		Four firms, case study
Ili et al. (2010)		Semi-structured, personal interview
Chiaroni et al. (2011)		Several firms, case study
Lichtenthaler (2010)		Several firms, case study
Spithoven et al. (2011)		Interview, questionnaires
Li et al. (2010)		Case study of Newzt
Li (2010)		Case Study of Tencent
Bianchi et al. (2010)		Interview, Financial DB
Gumus and Cubukcu (2011)	Survey; Turkish large enterprises	
Yang and Anderson (2011)		Case study of 18 companies,
Steiner et al. (2012) Yun and Mohan (2012) Yun and Mohan (2012)	Width of OI, depth of OI	One product, case study Twelve firms, half-structured questionnaire

Source: Yun et al. (2014)

of the subjectivity of survey-based results and the fact that it depends on questions developed from the Oslo Manual.

Yun et al. (2012) conducted a patent-based analysis of the development of more objective indices to measure the open innovation levels of firms. When firms independently apply for patents, it is most likely the result of closed innovation because the relevant technologies are the result of the independent R&D of firms. On the other hand, when firms apply for patents by going into partnerships with outside firms, universities, research institutes, or individuals, rather than independently, the same patent ideas can be considered to have been created as a result of open innovation outside of the firm, rather than within the firm. Accordingly, they defined the patents jointly applied for with external institutions or individuals as open innovation patents, and they defined the ratio of open innovation patents to the total number of patent applications as the open innovation level of a firm. Through this research, it was possible for them to demonstrate that the open innovation level of Hyundai Motors (approx. 9%) was far lower than that of Toyota Motors (approx. 30%), and the open innovation level of Samsung (approx. 13%) was also much lower than that of Nokia (approx. 65%). However, this study was limited in that it was not able to explain the levels of open innovation dimensionally during the patent application process of firms. For example, as a patent application by firms filed jointly with outside agencies is considered an open innovation patent, the measurement of the level of open innovation activity according to the number of outside agencies making a joint patent application, i.e., measurement of the intensity of open innovation, was not considered at all.

In summary, the concept of open innovation has been concretely established through case studies and surveys. The challenge has been to develop objective measures. Survey-based studies have been carried out despite the limitations inherent in such methods. A study by one of the authors attempted to define open innovation patents and to use this concept to measure open innovation levels more objectively. There are limitations in these newer approaches, i.e., measuring the intensity of open innovation, however. This study addresses all of these issues.

1.3.2 New Measure of Open Innovation

We can develop more diverse open innovation strategies and creative business models by enhancing the preciseness of the measurement of the open innovation level (McGill and Santoro 2009; Yun et al. 2010; Yun et al. 2016).

If we establish an objective measurement framework of open innovation by a patent analysis, as shown in Fig. 1.2, several advantages arise. The ratio of open innovation patents (ROI) refers to the ratio of open innovation patents collaboratively applied for with outside agencies over the total number of patent applications by certain firms, as shown in Fig. 1.2. It is considered an important index that shows the width of the open innovation of specific firms by measuring the proportion of the firm's innovation activities that are carried out in open innovation with outside agencies. ROI does not consider the number of patents applied for by a specific firm; rather, it reflects specific aspects or factors that reveal the open innovation behaviors or strategies of firms. ROI shows the breadth of open innovation because it indicates the breadth of the number of patents collaboratively applied for by firms against the total number of patents for which the firm applied.

This study defines a patent issued to more than two individuals, corporations, or institution applicants as an open innovation patent because the patent was jointly applied for with outside firms, schools, individuals, and/or other agencies. Thus, this can be considered as evidence that the patent technology was not developed from inside bodies but from collaborations with outside entities or acquired from outside the organization. Furthermore, there may be cases in which firms apply for patents jointly with outside organizations as a means of releasing their own technologies outside their own organization. Such cases are also defined as open innovation and can be measured through the proposed analysis framework.

However, a question may be raised as to whether a patent application made jointly by members inside a firm can be considered as an open innovation patent. However, a joint patent application by a firm with inside members is illegal given that it is an infringement of company interest; such patents are not considered to be open innovation patents. However, there may be cases of joint patent applications by firms and their inside members during the process of introducing idea developers

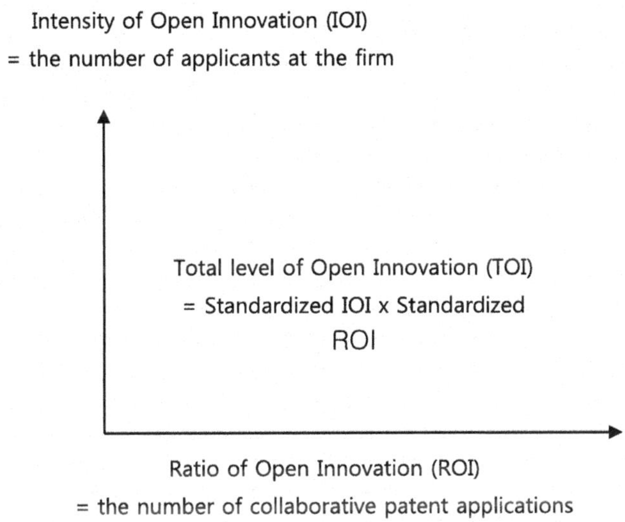

Fig. 1.2 The relationship among ROI, IOI, and TOI (Source: Yun et al. (2014))

from the outside to the inside by the firms or during the process of carrying out relevant patent-related business outside the firm by inside idea developers. These cases represent typical open innovation activities of firms and should be considered open innovation patents.

The intensity of open innovation (IOI) is an index that reflects the average number of applicants for all patents in any specific firm, as shown in Fig. 1.2. It is considered an important index that shows the number of outside agencies with which the firm is carrying out innovation activities. We should keep in mind that IOI differs from the average number of applicants for "collaborative patents," which refers to the intensity of open innovation in "collaborative patents." IOI shows us the total open innovation depth, while the intensity of open innovation in collaborative patents simply shows the depths of collaborative patents by a specific firm.

IOI should be considered to determine the depth of open innovation by firms. Moreover, IOI does not reflect the number of patents; rather, it reflects other aspects of patent applications, such as the average number of agencies applying for patents. IOI indicates the depth of open innovation because it shows the differences among firms in terms of the number of applicants out of all patents of individual firms.

Finally, the total level of open innovation (TOI) based on collaborative patents, as shown in Fig. 1.2, is identified by multiplying the standardized IOI by the standardized ROI. There are several reasons to develop the concept of the TOI. First, we need a unique concept to represent the OI of a firm when we analyze the OI effect in the firm. TOI shows us only one OI level of a firm. Second, we should keep in mind that IOI and ROI express different OI aspects of firms. TOI shows us different aspects of OI together based on the standardization of different measures and by multiplying them. Third, TOI should be based on the statistical methods used in determining IOI and ROI. If we use a statistical method, we easily obtain TOI from the IOI and ROI. The TOI concept does not have originality, similar to the IOI or

ROI concepts. When we need a unique level of OI for a firm, TOI can be the overriding concept.

Research Question

1. Select a case firm from a newspaper, and point out open innovation of the firm.
2. Select a case firm, and measure the degree of open innovation of the firm by your own method.

References

Alexander RD (1987) The biology of moral systems. Transaction Publishers, Piscataway

Arrow K (1962) Economic welfare and the allocation of resources for invention. The rate and direction of inventive activity: economic and social factors, Nber. UMI, Ann Arbor, pp 609–626

Auerswald PE, Branscomb LM (2003) Valleys of death and Darwinian seas: financing the invention to innovation transition in the United States. J Technol Transf 28(3–4):227–239

Axelrod R (1986) An evolutionary approach to norms. Am Polit Sci Rev 80(4):1095–1111

Axelrod R, Hamilton WD (1981) The evolution of cooperation. Science 211(4489):1390–1396

Bianchi M, Cavaliere A, Chiaroni D, Frattini F, Chiesa V (2010) Organisational modes for open innovation in the bio-pharmaceutical industry: an exploratory analysis. Technovation 31(1):22–33

Calia RC, Guerrini FM, Moura GL (2007) Innovation networks: from technological development to business model reconfiguration. Technovation 27(8):426–432

Cave J (1984) A formal analysis of the evolution of cooperation. RAND Corporation, Santa Monica

Chesbrough HW (2003) Open innovation: the new imperative for creating and profiting from technology. Harvard Business Press, Boston

Chesbrough H (2012) Why companies should have open business models. MIT Sloan Manag Rev 48(2):22–28

Chesbrough H, Vanhaverbeke W, West J (2006) Open innovation: a new paradigm for understanding industrial innovation. Open innovation: Researching a new paradigm. Oxford University Press, Cary, New York

Chiaroni D, Chiesa V, Frattini F (2010) Unravelling the process from closed to open innovation: evidence from mature, asset-intensive industries. R&D Manag 40(3):222–245

Chiaroni D, Chiesa V, Frattini F (2011) Open innovation journey: how firms dynamically implement the emerging innovation management paradigm. Technovation 31(1):34–43

Dosi G, Malerba F, Ramello GB, Silva F (2006) Information, appropriability, and the generation of innovative knowledge four decades after Arrow and Nelson: an introduction. Ind Corp Chang 15(6):891–901

Elberse A (2008) Should you invest in the long tail? Harv Bus Rev 86(7/8):88

Foray D, Lundvall B-A (1998) The knowledge-based economy: from the economics of knowledge to the learning economy. Econ Impact Knowl:115–121

Freeman C (2011) Technology, inequality and economic growth. Innov Dev 1(1):11–24

Gumus B, Cubukcu A (2011) Open innovation survey in top Turkish companies. Proceedings of PICMET technology management in the energy smart world, Portland, Oregon, 31 July–4 Aug, pp 1–6

Harinarayan V, Rajaraman A, Ranganathan A (2010) Hybrid machine/human computing arrangement: Google Patents

Hartman P, Bezos JP, Kaphan S, Spiegel J (1999) Method and system for placing a purchase order via a communications network: Google Patents

Howkins J (2002) The creative economy: how people make money from ideas. Penguin, London
Hughes B, Wareham J (2010) Knowledge arbitrage in global pharma: a synthetic view of absorptive capacity and open innovation. R&D Manag 40(3):324–343
Ili S, Albers A, Miller S (2010) Open innovation in the automotive industry. R&D Manag 40(3):246–255
Joshi N (1987) Evolution of cooperation by reciprocation within structured demes. J Genet 66(1):69–84
Kim SK (2009) Is open innovation a new innovation methodology?: a critical review on Chesbrough's open innovation theory. Technology Innovation Research, The Korean Society for Innovation Management & Economics, 99–133
Laherrere J, Sornette D (1998) Stretched exponential distributions in nature and economy: "fat tails" with characteristic scales. Eur Phys J B 2(4):525–539
Laursen K, Salter A (2006) Open for innovation: the role of openness in explaining innovation performance among UK manufacturing firms. Strateg Manag J 27(2):131–150
Leimar O, Hammerstein P (2001) Evolution of cooperation through indirect reciprocity. Proc R Soc London Ser B: Biol Sci 268(1468):745–753
Lee HH, Bae KS, Lee JW, In YY, Kwon TH, Lee WW (2016) Valuation method by regression analysis on real royalty -related data by using multiple input descriptors in royalty negotiations in Life Science area-focused on anticancer therapies. J Open Innov; Technology, Market, and Complexity 2(21):1–22
Levinthal D, Rerup C (2006) Crossing an apparent chasm: bridging mindful and less-mindful perspectives on organizational learning. Organ Sci 17(4):502–513
Li J (2010) Research on the mechanism of hi-tech enterprises' open innovation on the basis of knowledge chain. Proceedings of the IEEE international conference on advanced management science, Chengdu, 9–11 July, pp 163–167
Li Q, Yu X, Jin J, Chen J (2010) Open innovation in Cinese SMEs: a case study. In: Proceedings of the IEEE international conference on management of innovation and technology, Furama RiverFront, Singapore, 2–5 June, pp 726–730
Lichtenthaler U (2009) Outbound open innovation and its effect on firm performance: examining environmental influences. R&D Manag 39(4):317–330
Lichtenthaler U (2010) Technology exploitation in the context of open innovation: finding the right "job" for your technology. Technovation 30(7):429–435
Lotem A, Fishman MA, Stone L (1999) Evolution of cooperation between individuals. Nature 400(6741):226–227
Markusen A, Wassall GH, DeNatale D, Cohen R (2008) Defining the creative economy: industry and occupational approaches. Econ Dev Q 22(1):24–45
May J, Flint I (2009) Method and apparatus for facilitating online payment transactions in a network-based transaction facility using multiple payment instruments: Google Patents.
McDonough JF III (2006) Myth of the patent troll: an alternative view of the function of patent dealers in an idea economy. Emory Law J 56:189
McGill JP, Santoro MD (2009) Alliance portfolios and patent output: the case of biotechnology alliances. IEEE Trans Eng Manag 56(30):388–401
Moore GA (2002) Crossing the chasm: marketing and selling disruptive products to mainstream customers. HarperCollins, New York
Moran N (2007) Public sector seeks to bridge "valley of death". Nat Biotechnol 25(3):266–266
Nowak MA, Sigmund K (1998) Evolution of indirect reciprocity by image scoring. Nature 393(6685):573–577
Nowak MA, May RM, Sigmund K (1995) The arithmetics of mutual help. Sci Am 272(6):76–81
Porter ME (1985) Competitive advantage. Free Press, New York
Rai A, Reichman J, Uhlir P, Crossman C (2008) Pathways across the valley of death. Yale J Health Policy Law Ethics 8:1–36
Rapoport A (1965) Prisoner's dilemma: a study in conflict and cooperation. University of Michigan Press, Ann Arbor
Risch M (2012) Patent troll myths. Seton Hall Law Rev 42:457–500
Shove E (1998) Gaps, barriers and conceptual chasms: theories of technology transfer and energy in buildings. Energ Policy 26(15):1105–1112

Snow CP (1959) Two cultures. Science 130(3373):419–419
Spithoven A, Clarysee B, Knockaert M (2011) Building absorptive capacity to organise inbound open innovation in traditional industries. Technovation 30(2):130–141
Sroufe R, Curkovic S, Montabon F, Melnyk SA (2000) The new product design process and design for environment: "crossing the chasm". Int J Oper Prod Manag 20(2):267–291
Steiner A, Morel L, Camargo M (2012) Toward autonomy of ideas: conceptual framework of open innovation. Proceedings of the 18th international conference on engineering, technology and innovation, ICE, Munich, 18–20 June, pp 1–11
Stokes DE (1997) Pasteur's quadrant: basic science and technological innovation. Brookings Institution Press, Washington, DC
Taylor M (1976) Anarchy and cooperation. Wiley, London/New York
Teece DJ, Pisano G, Shuen A (1997) Dynamic capabilities and strategic management. Strateg Manag J 18(7):509–533
Trivers RL (1971) The evolution of reciprocal altruism. Q Rev Biol 46(1):35–57
Tushman M, O'Reilly C (2007) Research and relevance: implications of Pasteur's quadrant for doctoral programs and faculty development. Acad Manag J 50(4):769–774
Vad de Vrande V, De Jong JP, Vahaverbeke W, De Rochemont M (2009) Open innovation in SMEs: Trends, motives and management challenges. Technovation 29(6):423–437
Wedekind C, Milinski M (2000) Cooperation through image scoring in humans. Science 288(5467):850–852
West J, Gallagher S (2006) Challenges of open innovation: the paradox of firm investment in open-source software. R&D Manag 36(3):319–331
Yang J, Anderson T (2011) How open innovation strategy is reflected in the firms R&D efficiency DEA ranking?. Proceedings of the PICMET technology management in the energy smart world, Portland, Oregon, 31 July–4 Aug, pp 1–9
Yang JH, Yun JHJ (2014) New perspectives on open innovation from game theory: the Prisoners' dilemma and the ultimatum game. Asia Pac J Innov Entrep 8(1)
Yun JHJ (2010) Management of open innovation foundation policies. Republic of Korea, Book & World
Yun JHJ, Mohan AV (2012) Exploring open innovation approaches adopted by small and medium firms in emerging/growth industries: case studies from Daegu–Gyeongbuk region of South Korea. Int J Technol Policy Manage 12(1):1–19
Yun JHJ, Park SM (2012) Open innovation and performance of SMEs: comparison between Daegu/Gyeongbuk and other regions. J Bus Econ 28(1):1–22
Yun JHJ, Kim BT, Mohan AV (2009) Evaluating R&D management systems: strengths and weaknesses of universities and government-funded research institutes. Int J Technol Policy Manage 9(3):235–255
Yun JHJ, Park SM, Jung WY (2012) A study on the geographical boundary and location of open innovation. J Innov Clust 4(1):17–43
Yun JHJ, Won DK, Jung WY (2013). OCE: a new economic model of the dynamics of open innovation, applied to the smartphone sectors of Samsung, Apple, and Nokia. Proceeding of the R&D Management conference in Manchester
Yun JHJ, Won DK, Jung WY (2016) Dynamics from open innovation to evolutionary change. J Open Innov Technol Market Complex 2(7):1–22
Yun JHJ, Park S, Lim DW, Hahm SD (2010) Emergence of east Asian TFT-LCD clusters: a comparative analysis of the Samsung cluster in South Korea and the Chimei cluster in Taiwan. Asian J Technol Innov 18(1):201–228
Yun JJ, Avvari AV, Jeong ES, Lim DW (2014) Introduction of an objective model to measure open innovation and its application to the information technology convergence sector. Int J Technol Policy Manage 14(4):383–400
Yun JHJ, Won DK, Jeong ES, Park KB, Yang JH, Park JY (2016) The relationship between technology, business model, and market in autonomous car and intelligent robot industries. Technol Forecast Soc Chang 103(6):142–155
Zott C, Amit R (2008) The fit between product market strategy and business model: implications for firm performance. Strateg Manag J 29(1):1–26

Part II

Open Innovation in an Economic System and the National Innovation System

Dynamics of the Open Innovation Economy System

2

Abstract

The first purpose of this chapter is to understand the modern economy from the perspective of the technology and the market. The second and main purpose is to find ways to conquer the growth limits of capitalism by connecting and combining the technology and the market. Based on existing studies of the growth limits of capitalism by Marx and Schumpeter as well as recent discussions by Drucker, Rifkin, and Piketty, the Schumpeterian dynamic model of an open innovation economy system (OIES) is proposed as an answer to the research questions posed here.

OIES consists of an open innovation economy, a closed innovation economy, and a social innovation economy. The Schumpeterian dynamics of OIES arises from the positive interaction among the open innovation economy, closed innovation economy, and social innovation economy. The Schumpeterian dynamics of the OIES circle are from an open innovation economy, through a closed innovation economy and social innovation economy, and back to an open innovation economy again. In addition, the validation of the model for the Schumpeterian dynamics of OIES is improved by simulating the life cycle of the dynamics of OIES, the low-level OIES dynamics, and the high-level OIES dynamics and by inquiring about a practical economic system corresponding to each simulation situation. Next, through a comparative discussion between the linear steps of Schumpeter 1 and 2, the socialist democracy system, and the Schumpeterian dynamics of an open innovation economic system, the practical and theoretical characteristics of the Schumpeterian dynamics of OIES are clearly defined. Finally, the usage of the Schumpeterian dynamics model is proposed for connections and combinations between technology, market, and society.

This chapter is based on Yun J.H.J. (2015) "How do we conquer the growth limits of capitalism? Schumpeterian dynamics of open innovation economy system," *Journal of Open Innovation: Technology, Market, and Complexity,* 1(17), 1–20.

J.J. Yun, *Business Model Design Compass*, Management for Professionals,
DOI 10.1007/978-981-10-4128-0_2

Keywords
Schumpeterian dynamics of open innovation economy system

2.1 Introduction

2.1.1 Capitalism Has Arrived at Its Growth Limits

As of March of 2015, the growth of the global capitalistic economy appeared to have nearly halted. Specifically, the base rate of each nation's capital investment, which is the standard for capital income that serves as the key aspect of capitalism, was close to 0.00%. With this, the base rates of the capital investments of advanced nations practicing capitalism to some degree have also reached nearly 0%. In that the federal funds rate of the Federal Reserve Bank of the USA is 0.00–0.25%, the rate of the European Central Bank is 0.05%, and the rate of the Bank of Japan is 0.1%, evidence exists of a standstill. Except for China, India, and a few other countries, most major countries have arrived at an economic growth rate of nearly zero.

In particular, if the contributions of inflation and population growth are subtracted from the total economic growth, it can be observed that the economic growth rate of major capitalistic nations is close to 0%. Before the capitalistic economic system was introduced, a very low economic growth rate was recorded in the past. After the Industrial Revolution, early capitalistic markets recorded high economic growth rates (Piketty 2014, p. 25). However, after the 1980s, the USA, EU, and Japan recorded low economic growth rates of less than half of their respective rates over the previous 20 years. Although the quantity and distribution of technology and knowledge had increased globally and the world had become flat owing to global networks, the economic growth rates of major capitalistic nations decreased by more than half.

Since the mid-1990s, the global growth engine of capitalism has remained stagnant. The stagnation of economic growth is sometimes interpreted as a continuation of the aftermath of the global real estate bubble caused by the subprime mortgage crisis. However, the real estate bubble itself was caused by a structural distortion of capitalism stemming from the stagnation of the growth of capitalism during which the expected income rate of capital investment remains very low. Thus, it is not accurate to consider that the current stagnant growth of capitalism is part of the aftermath of the subprime mortgage crisis. A theoretical and practical discussion to define and interpret this stagnation of capitalism around the world is necessary to guarantee an alternative with which to handle the problem.

2.1.2 How Do We Conquer the Growth Limits of Capitalism?

This chapter starts with the recognition that capitalism has reached its growth limit. Indeed, it is considered as part of the business cycle, and it can be seen that the growth of global capitalism has not reached its limit but has been on the decline in the business cycle. Global capitalism has entered a period of stagnation. To overcome it, Piketty notes that the distribution structure should be improved as a new alternative. As part of this effort, he suggested a capital gains tax. His discussion corresponds to the growth limits of capitalism (Piketty 2014). Meanwhile, Ha-Sung Chang holds that as capitalism in Korea moves away from its original form, its soundness should be recovered through political intervention by the government. His logic handles the limits of capitalism at the national level (Ha-Sung Chang 2014). He also discusses the growth limits of capitalism through a list of 23 facts not widely known about capitalism (Chang 2002, p. 14; 2010, p. 11).

Since the financial crisis in the 1990s, the gross domestic product (GDP) per capita in Korea has remained around USD 20,000. However, the sales and profits of the top 10 or 30 companies that are listed are at their highest levels since the foundation of the country, and the current account surplus has continued to rise, reaching its highest level in March of 2015. Despite this good showing, and while growing, the Korean economy continues to show a relatively low rate of new job creation. The USA has achieved relatively high economic growth of 3–4% and new job creation based on quantitative easing. In most European countries and in Japan, economic growth has continued to slow down for decades.

The purpose of this chapter is to answer the research question, "How do we conquer the growth limits of capitalism?" In other words, the question of whether the growth of capitalism has reached its limit is not the subject of this study. However, to identify this clearly, the theory of the growth limits of capitalism is summarized. A discussion is presented on a perceptual basis with a theory model of development to overcome the growth limits of capitalism, which is the key element of this study.

2.2 Reasons Behind the Growth Limits of Capitalism: Findings from the Literature

The discussion of the growth limits of capitalism is mainly rooted in Marx's decreasing law of the surplus value of capital, the concepts of big business and socialist democracy of Schumpeter, and the general theory of employment, interest, and money by Keynes (Keynes 1935; Marx 1867; Schumpeter 1934, 1939, 1942). Adam Smith introduced the concept of equilibrium with regard to the capitalist economy and devised the self-purified market balance based on the price mechanism corresponding to the physical order of nature in theory. He formed the theoretical basis of the objection to growth limits of capitalism in the neoclassical school of thought (Marx 1867; Smith 1937). In fact, Marx said that a balance achieved between supply and demand in the market is a highly exceptional condition, holding that the market itself is inevitably imbalanced, as described in the Capital, Volume II (Marx 1978b).

However, with the rapid increase in the decision power of the economic growth of the technology and the market in the global economy in the late twentieth century and the twenty-first century, various discussions on the limits of capitalism have arisen. For example, there are the concepts of Drucker's post-capitalism, Rifkin's end of work and zero marginal cost, Piketty's acceleration of unbalanced distribution, and Ostrom's comedy of the commons (Drucker 1993; Ostrom 1990; Piketty 2014; Rifkin and Kruger 1996; Rifkin 2014; Rifkin and Kruger 1996).

2.2.1 Discussion in the Nineteenth Century and Early Twentieth Century

Marx discussed the labor process and the process of producing surplus value (Marx 1867, p. 197). According to his works, the absolute surplus value and the relative surplus value are generated from labor and are converted into capital. Thus, the general law of capitalist accumulation originates from the accumulation of surplus value, referred to as primitive accumulation (Marx 1978a, p. 784). Based on the logic of Marx, this tends to decrease surplus value (Marx et al. 1963). However, the tendency of decreasing interest in a capitalistic society was noticeable in the Western society after the Industrial Revolution in the nineteenth century and in the USA after the mid-twentieth century (Medio 1972; Wolff 1979). Marx predicted that capitalism will essentially reduce the ratio of surplus value, resulting in panic in capitalistic societies. The surplus value generated from labor is fundamentally converted into capital, and the accumulation of capital reduces its value. Lastly, there are financial panics in capitalist societies.

Schumpeter also addressed Marx's theory, i.e., that capital is accumulated through labor. He actively and sharply criticizes this theory (Schumpeter 1942, pp. 9–58). However, he also said, "Can capitalism survive? No. I do not think it can." He showed skepticism about the sustainability of capitalism in its current form at the time (Schumpeter 1942, p. 61), positing that the rate of increase in the total output has a certain limit. Schumpeter admits that the new combination based on the process of creative destruction as led by entrepreneurs was not an alternative to the continuous growth of capitalism. He insists that the new combination will disappear and that a closed season controlled by big business will emerge. Such a system will lead to the vanishing of investment opportunities (Schumpeter 1942, p. 111). However, with changes in the economic process brought about by innovation together with all their effects and the response they receive from the economic system, Schumpeter coined the phrase "economic evolution" (Schumpeter 1939, p. 83). This type of innovation combines factors in a new way or creates a new combination. However, the future of economic evolution as suggested by Schumpeter is the decomposition of capitalism and socialism, set with the model of the Labour Party of the UK (Schumpeter 1942, p. 167).

2.2.2 Discussion in the Late Twentieth Century and Early Twenty-first Century

Peter Drucker suggested that a knowledge-based society is a society wherein knowledge controls the main areas of society as well as the economy; he added that it is a form of post-capitalist society. First, knowledge brings about an Industrial Revolution integrated with working tools, manufacturing processes, and products. Second, knowledge leads to a productivity revolution, which significantly improves productivity. Third, knowledge leads to a management revolution, which is applied to it (Drucker 1994, p. 46). Drucker identified, even with a three-stage revolution, that the existing capitalist society is becoming a knowledge-based society, which is very different from the capitalist society. As a means of production, the importance of knowledge increases in a post-capitalist society. Thus, it is expected that, among the two important means of production in a capitalist society, the traditional role and the function of capital as well as the disappearance of labor will be redefined (Drucker 1994, p. 115). That is, labor that serves as an asset disappears, and an employees' society in which they serve as subjects is formed. Furthermore, because pension funds account for a major part of the capital, the prevalence of a capitalist society having no capitalists is increased. Drucker said that the size and sustainability of the productivity of a knowledge-based society depend on the effort to improve the productivity of knowledge and service employees, the improvement of knowledge productivity, the design of an organization based on knowledge, and the reliability of the goals of individuals. Of particular interest is that Drucker mentioned the necessity of the recovery of governmental and citizens' functions for the sustainability of a knowledge-based society (Drucker 1994, pp. 236, 251). This point is very similar to the vision of Schumpeter on social democracy. He called the knowledge-based society an entrepreneurial society, led by an innovation agent, who is an entrepreneur, and suggested that the requirement for society to thrive is substantial social innovation (Drucker 2014, p. 257).

Jeremy Rifkin identified that a world led by the progress of technological innovation would mean the end of work. For a growing number of working people who find themselves either underemployed or unemployed, the concept of trickle-down technology is of very little solace (Rifkin and Kruger 1996, p. 165). He holds that the characteristics of the new society are a decline of the global labor force, the reduction of the middle class, new divisions between high-tech winners and losers, an increasing rate of unemployment, and a more dangerous world. As a result, Ripken predicted that globalizing the social innovation economy would be realized by reengineering the work week, by new social contracts, and by empowering the third sector (Rifkin and Kruger 1996, p. 275). In this third Industrial Revolution, lateral power refers to transforming the energy, the economy, and the world. Ripken noted that a mutual cooperative, horizontal, and open economic system, different from the existing system, would be established and in fact is being established (Rifkin 2011, p. 277). That is the sharing economy, or the collaborative commons, summarized as the zero-marginal-cost society (Rifkin 2014, p. 7). This refutes the opinion of Hardin, who holds that in the end, all commons are destined to collapse in a head-on manner. It also completely coincides with the work of Rose and Ostrom, who once raised the question of the old idea, long held by economists, that individuals pursue only their self-interests in a market (Hardin 1968; Ostrom 1990; Rose 1986).

Piketty said that the capital-labor split in the twenty-first century occurs as the ratio of capital continuously increases the dynamics of the capital/income ratio around the world. He said that the acceleration of unequal distributions based on the capital of capitalism of the twenty-first century is a global phenomenon and suggested that a viable alternative to capitalism is a social state (Piketty and Goldhammer 2014, p. 471). A modern redistribution based on a social state does not include transferring income from the rich to the poor, at least not in such an explicit way. Instead, it includes financing public services and replacement incomes that are generally equal for everyone, especially in the areas of health, education, and pensions (Piketty and Goldhammer 2014, p. 479). All rich countries, without exception, faced the twentieth century in a form of equilibrium in which less than one-tenth of national incomes was consumed by taxes; however, they now face a new equilibrium in which this figure has increased to between one-third and one-half (Piketty and Goldhammer 2014, p. 476). However, it is true that social states in poor and emerging countries are different from those of rich countries in terms of status or tendency. A nation takes 10–15% of its national income in sub-Saharan Africa and South Asia; 15–20% in Latin America, North Africa, and China; and about 10% in India. However, it has been shown that the ratio of a nation's national income decreases, with this tendency accelerating the inequality between the capital and labor of nations (Piketty and Goldhammer 2014, p. 491). Piketty notes that overcoming the accumulated inequality of the distribution between capital and labor in capitalism by forming a social state, including a global capital tax, is an alternative to handle the stagnation of economic growth. This is also shown as a worsening of the distribution inequality of capitalism in the twenty-first century (Piketty 2010, p. 61).

The OECD found that the growing inequality of wealth decreases the growth rate of member nations. From this perspective, the growth limits of capitalism mean the skyrocketing of inequality of wealth. The inequality, unemployment and stagnant incomes, and globalization of the economy are currently occurring at once.

2.2.3 Discussion of the Growth Limits of Capitalism in Korea

Ha-Sung Chang identified that Korean capitalism is broken because it does not have the three growth elements, that is, employment growth, wage growth, and distribution growth. As an alternative, he suggests the concept of righteous capitalism (Ha-Sung Chang 2014, p. 521). For 5 years, from 2008 when the economic crisis occurred up to 2013, the average global growth rate of Korea was 1.7%, while the average growth rates of the USA and the UK were 0.8% and −0.6%, respectively. This led to capitalism in crisis. Chang pointed out that the crisis was caused by the structure of inequality and the worsening of inequality rather than by the slowdown of growth, also saying that the fruits of growth did not have an impact on people's lives (Ha-Sung Chang 2014, p. 20). He noted, over the past 30 years when market fundamentalism dominated, that the conditions of employment were aggravated, with unstable employment increasing such occurrences as an income inequality, class polarization, economic growth structure without employment, and increased numbers of low-wage employees and temporary workers. Korea has experienced the same economic conditions. Ha-Sung Chang found that as corporate income continuously increases against household income among household income, corporate income, and government income, which together comprise the gross national income of a country, the internal reserves of the listed companies rise without end, resulting in a lack of corporate investment for job creation, a decrease in consumption in the household section, and the stagnant growth of capitalism in Korea. He takes an optimistic view that the growth limits of capitalism in Korea can be overcome by converting capitalism to righteous capitalism guaranteed by a competing system, a fair market, and the realization of justice. He added that the current stagnation in growth can be handled. In particular, he identified that based on trust in shareholders, who are the owners of assets in modern capitalism rather than capitalists, if shareholder capitalism in the form of the investment of foreign companies in Korea and even in the state-run firms of state capitalism as in China and Russia is based on competition, fairness, and justice, capitalism, even with its growth stopped, will eventually see growth again (Chang 2014, p. 247).

The discussion of partnered growth between the former Prime Minister of Korea Un-Chan Chung and Professor Jang-Woo Lee also showed that as the growth of Korean capitalism is halted, it is necessary to have capitalism with partnered growth as an alternative to ensure sustainable growth. This shows that Korean capitalism does not grow or that its growth is significantly stagnated (Jang-Woo Lee 2011, p. 49). Chung mentioned that if there are strong intentions by the government to cause changes in large companies with the self-help from small- and medium-sized enterprises (SMEs) for partnered growth, crucial for the new capitalism, the alleviation of polarization, improvements in job stability, and the continuous creation of new jobs can be achieved. These three factors should be addressed with regard to capitalism in Korea. He said that the key values of partnered growth, including profit sharing, are not applied to companies and the economy but to both the philosophy of life and the value of a new social community. Shared value creation focuses on identifying and expanding the connections between societal and economic progress (Porter and Kramer 2011). Shared value as the next evolution of capitalism will not only hold key value, unlocking the next wave of business innovation and growth, but will also reconnect the success of companies and communities in ways that have been lost in an age of narrow management approaches, short-term thinking, and deepening divides among institutions in society (Porter and Kramer 2011). Lee holds that the new coexistence of large-, small-, and medium-sized companies through partnered growth, a rule of the game that has no loser, is a means of building a sustainable economy (Jang-Woo Lee 2011, p. 15). However, he suggested the corporate ecosystem theory for the theoretical background of partnered growth, referring to the basic theory that creates a cooperative system among companies and the theory of behavioral change for partnered growth to focus on the sustainable development of the Korean economy (Jang-Woo Lee 2011, p. 111). Lee also noted that partnered growth is necessary to overcome the weakness of the capitalist economic system, but his view is different from those of Porter and Un-Chan Chung, who both reported that partnered growth is an alternative to overcome the growth limits of capitalism in its current form. In an era when great ideas can sprout from any corner of the world and when IT has dramatically reduced the cost of accessing them, it is now a conventional wisdom that virtually no company should innovate on its own (Pisano and Verganti 2008). In this regard, Lee insisted that partnered growth is essential for the survival and development of large companies as well as start-ups and SMEs (Jang-Woo Lee 2011, p. 190).

Lee noted that Korea is in a seriously unequal economic condition which is being made worse (Lee 2015, pp. 491–495). According to him, the growth limits of the Korean economy include extreme inequality based on several aspects, such as a development strategy of being an export-based economy, a large-scale urgent migration from rural areas to urban zones, a big business-based national economic system, and the family management system of big businesses.

2.3 Economy Model to Conquer the Growth Limit of Capitalism

2.3.1 Open Innovation Economic System with a Good Balance Between Three Subeconomies

The open innovation economic system (OIES) is a macroscopic economic system wherein a subeconomic system based on open innovation led by start-ups and small and medium enterprises (SMEs), a subeconomic system based on closed innovation led by big businesses, and a subeconomic system based on social innovation economy led by social enterprises or independent third sectors are well interconnected, thereby affecting each other. OIES essentially targets the economic system of one nation. However, an identical macroeconomic system could be applied to both global and regional economic systems.

That is, the capitalist economic system in a modern society is basically composed of the three subeconomic systems described in Fig. 2.1 regardless of the characteristics of the operable political system. An open innovation economy indicates an economy based on SMEs or start-ups led by individual entrepreneurs. It features a new combination between the technology and the market, as suggested by Schumpeter (Brunswicker and van de Vrande 2014; Schumpeter 1934, pp. 15, 65). Open innovation economy is characterized in such a way that the original producer of technology, including knowledge, is not the same as the subject delivering the production to a market (Chesbrough 2003, p. 43). Thus, the open business model, which is a new combination between a technology and a market and which led by

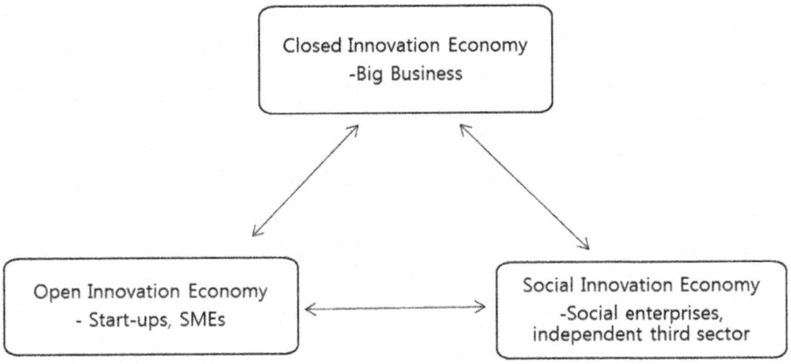

Fig. 2.1 Components and structure of an open innovation economic system (Source: Yun (2015))

various entrepreneurs, defines the growth and development of an open innovation economy (Chesbrough 2007, 2010, 2013, p. 2). A closed innovation economy is led by mainly big businesses, wherein monopolistic practices are dominant and investment opportunities no longer exist (Schumpeter 1942, pp. 81, 87). In this case, large companies create their own value based on the technology they have accumulated internally and transfer it to a market. Thus, they lead the closed innovation economy (Chesbrough 2003, pp. 21–24). A social innovation economy indicates that in the economy, specific technology or knowledge creates a social value that meets social requirements without the intermediation of the market and provides it to the society (Rifkin and Kruger 1996). Social innovation refers to innovative activities and services that are motivated by the goal of meeting a social need and that are predominantly diffused through organizations whose primary purposes are social (Mulgan 2006). The so-called sharing economy, exemplified by Airbnb and Uber, and the collaborative economy, represented by open source communities, are concrete examples of a social innovation economy which connects new technology and social requirements (Belk 2014; Kostakis and Bauwens 2014; Zervas et al. 2014). Social enterprises that act in a social innovation economy access the Internet of Things through a plug. They also utilize and interact with open and distributed types of architecture to create peer-to-peer horizontal collaborative commons (Rifkin 2014, p. 109; Zervas et al. 2014). With the activation of the Internet of Things, including the communication Internet, the logistics Internet, and the energy Internet, productivity significantly increases, and a zero-marginal-cost society is realized. Thus, the implementation of a sharing economy is expected to reorganize most aspects, such as energy, residences, and automobile logistics, based on access through ownership (Rifkin 2014, p. 389; Sundararajan 2013; Weitzman 1985).

However, for OIES, the dynamics of the three subeconomic systems are cyclical, as shown in Fig. 2.2. An economic system in which the dynamics of OIES actively occur can evolve and continuously create new jobs without experiencing the stagnation of growth. Thus, the growth limits of capitalism can be overcome when the dynamics of OIES actively occur based on the interconnections between the three subeconomic systems. In this case, creative start-ups and SMEs of the open innovation economy are initially provided to the closed innovation economy in open and

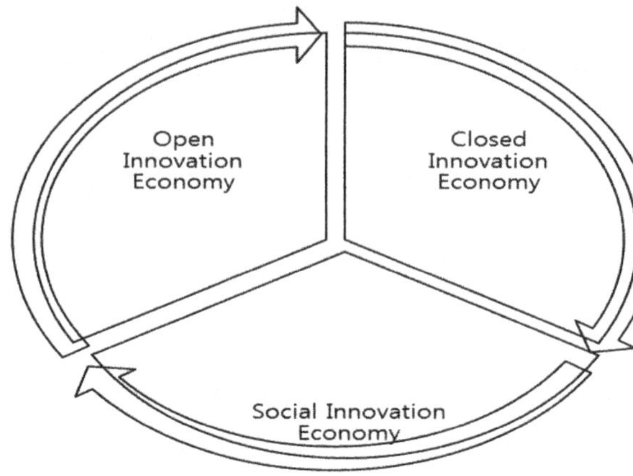

Fig. 2.2 Dynamics of an open innovation economy system (Source: Yun (2015))

innovative ways, such as M&As, technology licensing, or open platforms such that big businesses can continuously create new combinations in a short period of time for continuous job creation. Second, virtuous large businesses of the closed innovation economy distribute wages and take on a sufficient tax burden through large-scale employment. They continuously support the social innovation economy in an indirect way through this or in a direct way through voluntary contributions. Third, the social innovation economy plays a decisive role in the formation and development of a market by nurturing the social enterprises and social values created through the creative and newly operable combinations between technology and society. That is, the social innovation economy actively provides the seed for open innovation, which is a new combination between the creative technology and the market in the dynamics of OIES.

2.3.2 Three Interactive Relationships in Open Innovation Economy System

Companies in the open innovation economy are transferred to the closed innovation economy through mergers and acquisitions (M&A), and they gain the opportunity to mass-produce new business models for a short period of time. Large companies in the closed innovation economy can easily and rapidly undertake new combinations between the technology and the market through corporate venture capital (CVC) investments. Various virtuous interconnected relationships, including the M&As discussed in Fig. 2.3, are important factors for the activation of the dynamics of an open innovation economic system. For example, Apple bought approximately 20 technical companies related to smartphones through M&As and opened the App Store to realize partnerships with many SMEs around the world in their effort to enter into the smartphone industry in a short time, thus creating millions of employment opportunities. Google also entered into the smartphone industry over a short time through M&As with ten SMEs and start-ups, including the company which created the Android OS, to create more new jobs around the world, especially in the

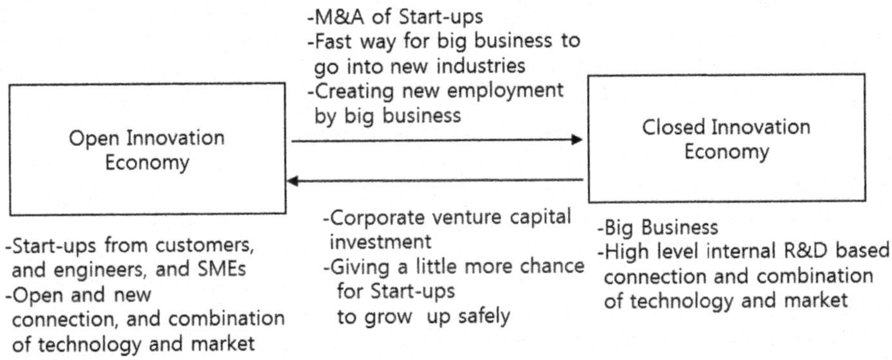

Fig. 2.3 Concrete relationships between the open innovation economy and the closed innovation economy (Source: Yun (2015))

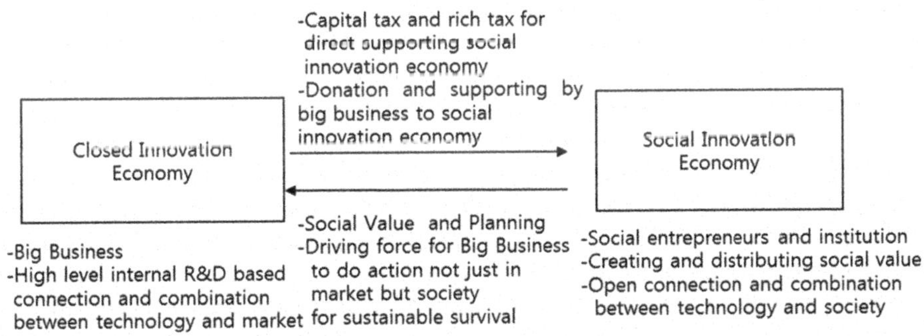

Fig. 2.4 Concrete relationships between the closed innovation economy and the social innovation economy (Source: Yun (2015))

USA. IBM became a software company from a manufacturing company through positive M&As with promising SMEs and start-ups. It achieved a new combination between the technology and the market which is different from those in the existing computer industry and created new jobs that replaced many in the presently declining industry. SMEs and the large companies of Silicon Valley in the USA that joined the Corporate Business Model Innovation Program of UC Berkeley confirmed the pursuit of mutual interests based on the virtuous interactive relationships described in Fig. 2.3. Large multinational corporations (MNCs) in the closed innovation economy basically apply various open innovation strategies for relationships between SMEs and start-ups (Mortara and Minshall 2014).

The closed innovation economy and the social innovation economy described in Fig. 2.4 form the core of corporate social responsibility (CSR). In particular, in terms of the relationships between a big business and the local society as well as between a big business and government, large companies should consider social and environmental interests and voluntarily cooperate with stakeholders for their sustainability and survival (Crowther and Aras 2008). Big businesses in a closed

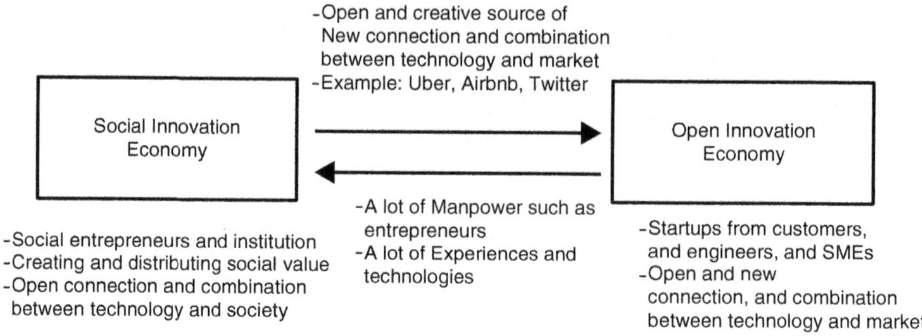

Fig. 2.5 Concrete relationships between a social innovation economy and an open innovation economy (Source: Yun (2015))

innovation economy directly and indirectly support the social innovation economy. Through this process, a big business directly earns a social reputation, which is essential for its long-term survival, while it also indirectly benefits from the final marketization of the value created in an open innovation economy by social enterprise. The relationship described in Fig. 2.4 offers the direct benefit of securing potential customers to large companies and contributes to the various and continuous production of social value in a social innovation economy. In addition, it continuously allows new combinations between the technology and the market, which are essential for the survival of large companies. Increasing numbers of companies known for their hard-nosed approaches to businesses, such as Google, IBM, Intel, Johnson & Johnson, Nestle, Unilever, and Walmart, have begun to embark on important shared value initiatives (Porter and Kramer 2011). The simple macroeconomics of profit-sharing possesses natural immunity to stagnation (Weitzman 1985) (Fig. 2.5).

In a social innovation economy, social enterprises create social values by combining technology with the society, which becomes a source of new combinations between the creative technologies and the market. Open social innovation functionally defines open combinations as those which are between technology and society. Open social innovation (OSI) is the application of either inbound or outbound open innovation strategies, along with innovations in the associated business model of the organization, to social challenges (Chesbrough and De Minin 2014). Meanwhile, many SMEs and start-ups in an open innovation economy attempt to join the social innovation economy with their experience and know-how and become a major supplier of knowledge and manpower for the social innovation economy. In the open innovation economy, the creative source of new or shifting start-ups through the new combination of technology and the market is based on the social innovation economy. In addition, even if the social innovation economy is financially supported by the closed economy, the actual manpower, know-how, and experience are supported by the open innovation economy. The case of the Eighth National Biennial Grassroots Innovation Awards of India, as identified through participant observation, is very similar. All 41 winners, including the three student award winners, were supported with manpower, patent applications, product development, and market sales know-how beyond the social type from the open innovation economy, as exemplified by the Honey Bee Network. Honey Bee Network is a crucible of like-minded individuals, innovators, farmers, scholars, academicians, policy-makers, entrepreneurs, and nongovernmental organizations (NGOs) in India. They were also financially supported by a closed innovation economy, in this case the National Innovation Foundation of India. This action has vitalized the social innovation economy. If both economies configure the innovation community with a flat and open collaboration network, the most active collaboration can be realized. In the case of India, this role is played by the Honey Bee Network (Pisano and Verganti 2008).

2.3.3 Theoretical Validation of the Dynamics of Open Innovation Economy System

First, the theoretical basis for the relationships among the open innovation economy, its closed innovation, and the social innovation economy can be found in the innovation and economy development through the new combinations of Schumpeter as well as in the open innovation for the open connection between the technology and the market and the open business model for the open combination of Chesbrough. To produce other items or identical items using a different method means to combine these materials and forces in various ways. This is known as a "new combination" (Schumpeter 1934, p. 65). Innovation combines factors in a new way or carries out new combinations (Schumpeter 1939, p. 84). In addition, entrepreneurs carry out innovations (Schumpeter 1939, p. 100). Open innovation means that valuable ideas can come to a company and the market both internally or externally (Chesbrough 2003, p. 43). In addition, a business model serves as an intermediate construct that links those technical and economic domains (Chesbrough 2003, p. 69). The knowledge economy, the open connection during open innovation and the open combination in the open business model of technology, and the market are the driving forces of new start-ups and creative value creation (Chesbrough 2010).

Second, the theoretical framework of large businesses based on the closed innovation economy is Schumpeter's monopolistic practices, closed season, and corporate social responsibility, as well as the theory of partnered growth by Un-Chan Chung and Jang-Woo Lee. If large businesses continuously pursue monopolistic behaviors and closed strategies, investment opportunities will vanish (Schumpeter 1942, pp. 87, 111). In addition, if large companies directly and indirectly support the social innovation economy and concept of creating shared value (CSV), it is essential to guarantee corporate social responsibility (CSR) for long-term survival (Holme and Watts 1999; Porter and Kramer 2011). The theory of partnered growth, linked to the win–win growth strategy and the partnered growth between SMEs that represents the open innovation economy and large businesses that represent the closed innovation economy, also serves as the theoretical basis for a mutual, virtuous, and cyclical relationship between the open innovation economy and the closed innovation economy (Jang-Woo Lee 2011, p. 15).

Third, the theoretical basis of the social innovation economy is Schumpeter's socialist blueprint, Ostrom's governing the commons, and the sharing economy, or CSV. Socialist management can start from a system of values that has evolved owing to their capitalist predecessors (Schumpeter 1942, p. 172). With the development of the Internet, the marginal cost approaches zero, and the sharing social innovation economy emerges in the automobile, home, and energy fields. The social blueprint of Schumpeter is partially realized. Lastly, the shared economy offers a happier life, more money, a more flexible lifestyle, a reduced reliance on debt, and more trust in strangers as the subjects of the sharing economy that creates sharable value essentially to give benefits to its participants (Benkler 2004). According to Ostrom, common-pool resources (CPR) can be managed successfully without falling as prey to the "tragedy of the commons" through the design of durable, cooperative institutions that are organized and governed by the resource users (Ostrom 1990, p. 25). Social open innovation based on an open combination between technology and society receives direct and indirect support from big businesses based on CSR and CSV. In addition, the interaction between the social innovation economy and open innovation is in fact based on CSV.

Fourth, the dynamics of OIES has various theoretical bases and as follows:

1. The theoretical basis of the dynamics of OIES comes from Schumpeter. The dynamics of OIES basically complements the one-way, three-step discussion of dynamics, akin to Schumpeter's individual entrepreneur-based new combination, big business-based monopolistic practices, and socialist blueprint based on socialist democracy by changing it to simultaneous and feedback loop-style dynamics (Schumpeter 1934, p. 59; 1939, pp. 65–106; 1942, p. 87, 172, 232).
2. Simon's bounded rationality defines organizational learning beyond personal recognition and offers a theoretical basis for the dynamics of OIES as well as his organizational dynamics (Simon 1982, 1991).
3. The discussion of dynamics capability in a company suggested by Teece is the basis of the dynamics of OIES centered on learning dynamics at the basic organizational level (Teece et al. 1997).
4. Christensen's discussion to find the industrial dynamics based on the viewpoint of open innovation and Kong Rae Lee's research inquiring into the innovation dynamics of the Japanese machine tool industry, with users collectively forming the theoretical basis of OIES dynamics because they analyze open innovation based on economic dynamics (Lee 1996). Industrial dynamics must increasingly be conceived in terms of convergence and divergence rather than industry-bounded trajectories by open and industry-transcending patterns of innovation (Christensen 2014; Christensen et al. 2005).
5. Linsu Kim's theory, which suggests organizational learning through catch-up growth and the dynamics of Korea's technological learning through innovative imitation, is also an important theoretical basis of the dynamics of OIES (Kim 1997, 1998).
6. Keun Lee's and Chaisung Lim's discussions of the learning and dynamics of the Korean industry through a catch-up strategy are the basis of the dynamics of

open innovation (Lee and Lim 2001). They introduce the external base by adding knowledge-based learning and industrial dynamics through catch-up growth with an internal base and by explicitly applying the open innovation-based industrial dynamics to the catch-up model.

7. For the effect of the open innovation policy on the National Innovation System (NIS), the inquiry into the dynamics of open innovation at the national level through system dynamics is also matched with the theory of the dynamics of OIES (Yun et al. 2015).

2.4 Simulation of the Dynamics of Open Innovation Economy System

2.4.1 Natural Life Cycle

In the twenty-first century, knowledge-based economies or knowledge-based societies in which knowledge production and distribution are activated around the world and knowledge and technology become a major determining factor are already established as dominant characteristics (Burton-Jones 2001; Foray and Lundvall 1998). OIES dynamics and a knowledge-based economy can be simulated through a mental experiment, as shown in Fig. 2.6.

That is, if the ratio of the social innovation economy is 100%, the ratio of the open innovation economy grows in the early stage. The closed innovation economy, based on big businesses, then develops. However, as OIES dynamics evolve, the ratio of the open innovation economy significantly exceeds the majority of the social innovation economy.

In addition, in the middle stage, the open innovation economy accounts for the largest portion, followed by the social innovation economy and the closed innovation economy. The open innovation economy actively operates, and the social innovation economy continuously decreases. The closed innovation economy, based on large businesses, slowly increases. The OIES dynamics in the middle stage accelerates to increase the speed of economic growth and employment.

If the OIES dynamics matures when the social agreement and government intervention are weak, the economy, which is led by the closed economy based on large businesses, as in the example shown in Fig. 2.6, is established. In addition, the ratio of the social innovation economy decreases, to a point much lower than the majority, while that of the open innovation economy also decreases to the point of the majority or lower than that of the majority. The maturity led by the closed innovation economy restricts new employment due to the limited new combinations. The ratio of the open innovation economy and the social innovation economy, accounting for the largest portion of job creation, significantly decreases, thereby resulting in a high unemployment rate. By any chance, if economic growth led by the closed innovation economy lasts for a certain period, growth continues without employment.

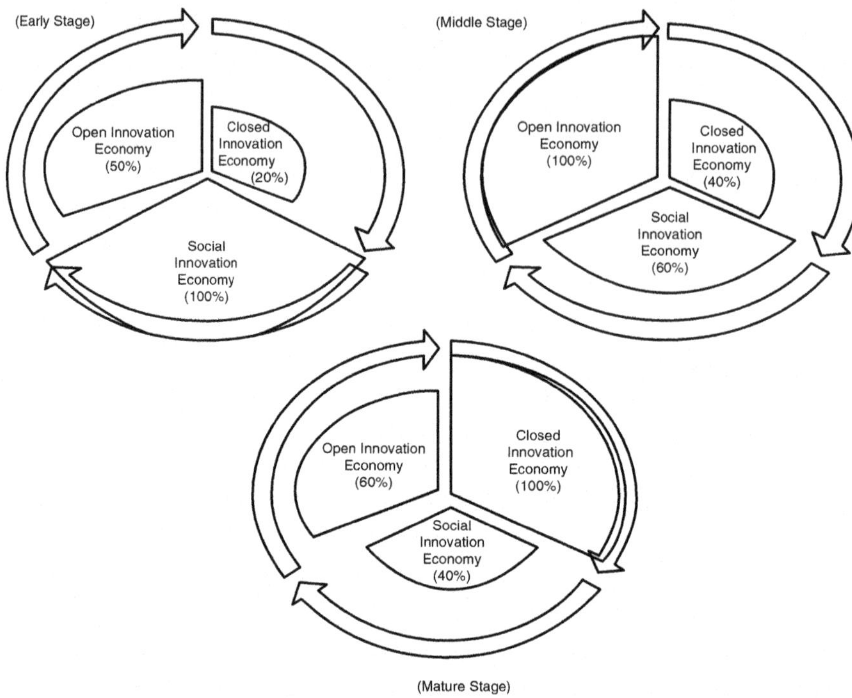

Fig. 2.6 Example of the natural life cycle of OIES (Source: Yun (2015))

2.4.2 Low-Speed and High-Speed Cases

The low speed of the dynamics of OIES as well as the economic growth and high unemployment rate are shown in Fig. 2.7a and b.

In Fig. 2.7a, there is a strong open innovation ecosystem based in SMEs as well as a sound and sizable closed innovation economy but a weak social innovation economy. Hence, growth stagnates. In this case, the financial support to the social innovation economy by large businesses is weak, and the support to the social innovation economy becomes insufficient in terms of experience, know-how, and manpower from the open innovation economy. The subjects of the open innovation economy focus on limited open innovation in which the combination of the current technology and market is newly improved rather than on the creation of new jobs through creative and new start-ups based on the seed grown in the social innovation economy. In addition, there are too few open innovation relationships between the open innovation economy and the closed innovation economy. The closed innovation economy, based on large businesses, also creates new sources of technology inside and focuses in the activity of delivering the technology to the existing or expanding market, resulting in a more aggravated situation of no additional job creation. Japan in the 2000s is similar to the situation depicted in Fig. 2.7a. Without the effort to vitalize the social innovation economy and with an open innovation

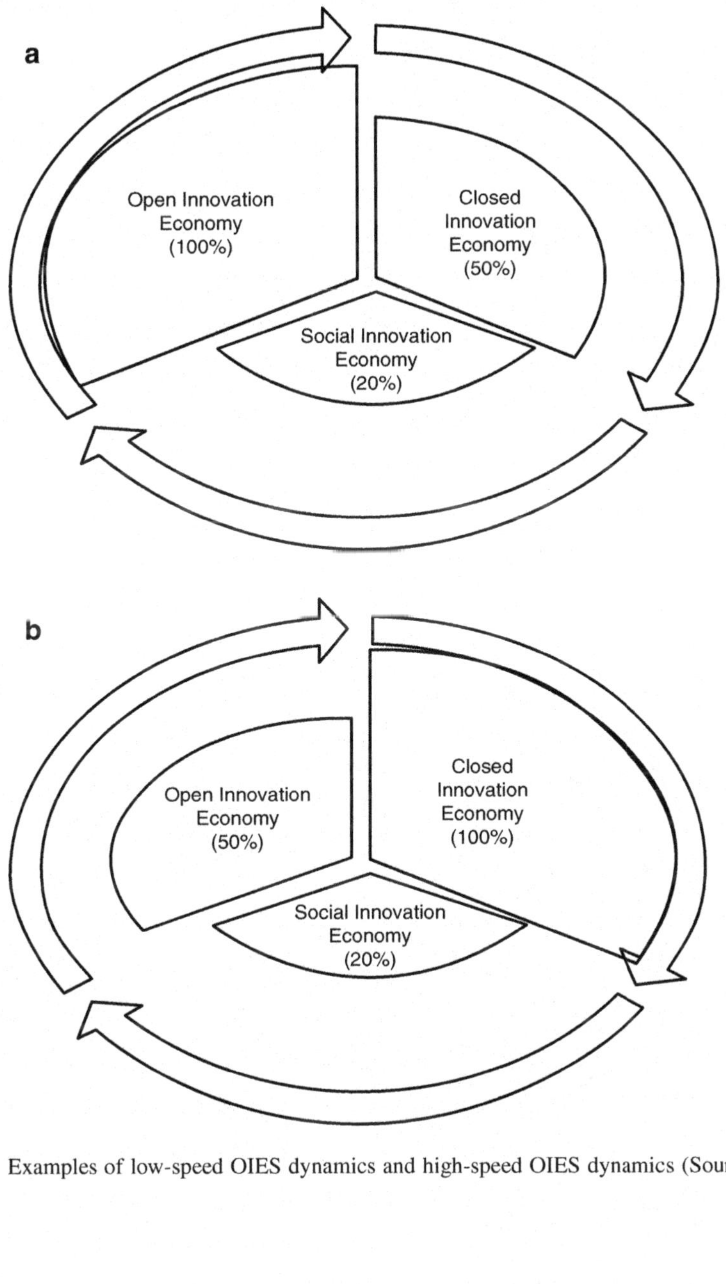

Fig. 2.7 Examples of low-speed OIES dynamics and high-speed OIES dynamics (Source: (Yun 2015))

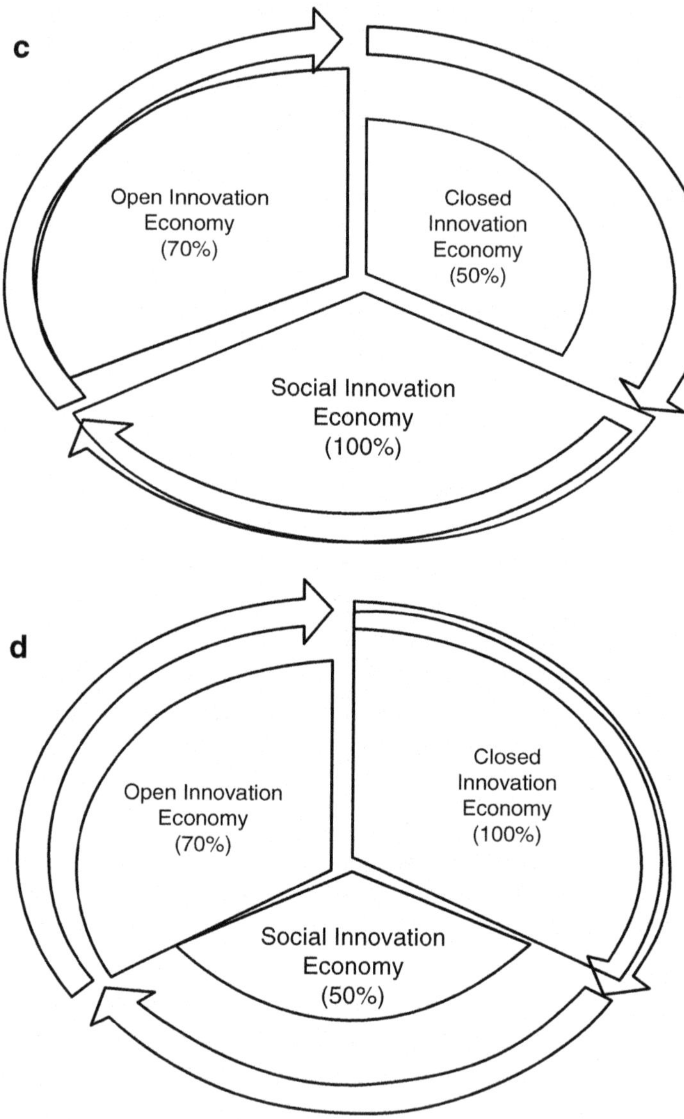

Fig. 2.7 (continued)

strategy that based the relationship between the open innovation economy and the closed innovation economy, it is inevitable that this economic system promotes growth without job creation, resulting in growth stagnation.

In Fig. 2.7b, large businesses sustain closed innovation-based growth without a social contract or national intervention, and SMEs and start-ups that are competing in an open innovation economy are depleted. Of course, the social innovation economy with slight direct and indirect support, in the good will of the closed innovation economy, is more contracted to a size that decreased during the OIES dynamics procedure. In this case, big businesses do not prefer to cooperate with SMEs or start-ups through open innovation strategies such as friendly M&As, partnerships, and technology-licensing agreements because they are based on their strong market shares and capital power. In Fig. 2.7b, economic growth stagnates because it is led by large businesses, and even if economic growth is rapid owing to environmental factors, it leads to growth without employment and thus stops economic growth. Figure 2.7b depicts a situation similar to that of Korea as well as most European nations apart from Germany.

It should be noted that the common point of the two case models in which the dynamics of OIES are slow is a weak social innovation economy. The social innovation economy is a necessary condition that determines the vitality of the dynamics of OIES.

The rapid speed of the dynamics of OIES, the low unemployment, and the rapid speed of economic growth are shown in Fig. 2.7c and d. In Fig. 2.7c, social enterprises and new social entrepreneurs in a sizeable and strong social innovation economy actively combine technology and society in various ways. To do this, massive support from the government or strong financial assistance from large companies should be established. In addition, in an open innovation economy, manpower and know-how are massively provided to activate the social innovation economy and convert new combinations between technology and society to new combinations between the technology and the market so as to establish the foundation of start-ups for the open innovation economy. Nations at an early stage of economic development are examples of this case. Thus far, only China's economy has closely resembled this case, although now, India's economy also applies. The Indian government has made efforts to strengthen the open innovation economy, vitalize the social innovation economy, support the social innovation economy financially, and promote the growth of large businesses at a certain scale by linking the open innovation economy and the social innovation economy through the Grassroots Innovation Awards, the Innovation Festival of India, and the Innovation Foundation of India, respectively. Korea's rapid economic growth until the 1980s exemplified this case.

In case of Fig. 2.7d, this economic system is formed through a social contract or through strong government intervention when the growth of the OIES dynamics is stagnated. The mutual development of big business with SMEs or start-ups is pursued through strong regulation by the government of the closed innovation system and of open innovation strategies such as friendly M&As, partnerships, and technology-licensing agreements. For example, the USA continuously implements and develops the world's first and strongest regulation system of big businesses in the closed innovation economy, such as stipulating unfair businesses as unlawful and allowing people to take legal action through the Sherman Act of 1890, the Clayton Act of 1914, the Hart–Scott–Rodino Act of 1976, and the Federal Trade Commission. In addition, large businesses directly provide financial support, or the nation activates various types of social economies through taxes. In such a case, a social innovation economy continuously creates jobs in all economic systems. In the short term, it becomes the source of a new combination between the technology and the market of SMEs and start-ups. In the long term, it becomes the seed of continuous new combinations of large businesses. In the USA, the image of the best large companies set by Rockefeller, Carnegie, and others in the 1910s has led to continuous contributions and the financial support of large businesses to the social innovation economy. The economic systems of the USA and of European countries from the 1970s to the 1980s, enjoying rapid economic growth, are examples of this case. The economy of the USA with a high economic growth rate of 4% and little promotion of job creation as of late 2014 exemplifies this case. President Obama suggested the "Startup America Partnership" program to activate a mutual virtuous circle between large companies and SMEs, with the ratio of the open innovation economy to the closed innovation economy expected to increase. This causes a rise in the economic growth rate.

2.5 Schumpeterian Dynamics

2.5.1 Schumpeterian Dynamics of Open Innovation Economy System

It is clear that the dynamics of OIES have theoretical and practical implications when comparing them with the dynamics of economic evolution and development as discussed by Schumpeter.

Although Schumpeter did not concretely mention the technology and the market, he is the first scholar who clearly suggested economic evolution and dynamics while explaining that the new combination between the technology and the market promotes innovation, creates new jobs, and develops the economy (Schumpeter 1934, p. 15). However, the economic evolution or development discussed by Schumpeter is linear. Schumpeter separates the subject that leads to the new combination in the first stage from the capital class, defining it as an entrepreneur.

Schumpeter suggested that the subject of economic evolution and the development of the second stage are focused on large businesses. A large business attempts to increase its rates of total output, to create new combinations through massive investments in research and development, and leads the process of creative destruction (Schumpeter 1942, p. 81). However, the large business-based economy goes through monopolistic practices, closed seasons, and disappearance of investment opportunities. As an alternative to overcome the limits of the second stage, Schumpeter introduced socialist blueprints that focus on the civilization of capitalism, providing the Labour Party in the UK as an example.

The three stages of the logic of economic development of Schumpeter's are generally accepted as the types of subeconomies in the dynamics of OIES. Here, we refer to this as the Schumpeterian dynamics of OIES. The linear logic of economic development in Schumpeter's theory has been practically complemented based on the practical experience with economic development and academic performance accumulated over the past 65 years since the death of Schumpeter. Thus, the Schumpeterian dynamics of OIES were created.

All capitalist economic systems have a development stage identical to the economic development stage model of Schumpeter. However, the reasons why the current situation differs from the final expectation of Schumpeter and why the economic development speed and employment of capitalistic economic systems differ can be gleaned from the fact that economic development during the three stages is not performed on a step-by-step basis but overlaps in a circular manner. These overlapped economic development dynamics are shown in various ways in accordance with each economic situation or with external conditions.

Therefore, in the Schumpeterian dynamic of OIES, the distinction between low and high dynamics is not absolute. The models and practical cases shown in the simulation are merely a few examples. However, it is essential to introduce a new combination between the technology and the market for the creative and new emergence of SMEs and start-ups and a new combination between technology and society for the development of creative social enterprises and the social innovation economy, as well as a proper open innovation strategy to maintain the virtuous cyclic relationship between the closed innovation economy and the open innovation economy as well as with the social innovation economy and to sustain creative and new combinations.

2.5.2 New Combinations as the Core in Schumpeterian Dynamics

The general conditions under which to invigorate the Schumpeterian dynamics of OIES include five items. First, the new combination between the technology and the market should be activated to encourage the emergence of new creative SMEs and start-ups. Second, large businesses should continuously make efforts to introduce new business models and create new jobs by creating new combinations over the short term through friendly M&As, partnerships, and technology-licensing agreements with SMEs and start-ups rather than simply focusing on its internal R&D. Third, large businesses should make active contributions to vitalize the social innovation economy through direct donations and should also act indirectly through the payments of sufficient taxes for creative new combinations between the technology and the market over the long term. Fourth, the social innovation economy should actively create social values that are necessary for the society through the social open innovation of new combinations between technology and society, instead of only focusing on the production of social values. Due to the third Industrial Revolution, mainly based on the Internet, various shared values based on zero marginal cost have sharply increased, and the importance of social innovation economy has increased as well. Fifth, SMEs and start-ups should actively provide manpower, technology, knowledge, and experience to the social innovation economy based on their expertise in creating new combinations, as this is the source of creative and new combinations between the technology and the market.

2.5.3 Four Agencies in Schumpeterian Dynamics

The first agencies are entrepreneurs from start-ups or SMEs in an open innovation economy. They should not hesitate to connect and combine the technology and the market creatively and openly. Their entrepreneurship is the fundamental source of the sustainable development of the economy in terms of quantity and quality.

The second agencies are large businesses. They create substantial increases in employment and quantifiable economy development. They invest in internal research and development and in market creation by themselves. However, they should also choose open innovation strategies such as M&As, good partnerships, or technology-licensing agreements to build new business models rapidly and continuously. Moreover, they should not avoid donating in the social innovation economy and paying sufficient taxes for the maintenance and development of Schumpeterian dynamics.

The third agencies are social entrepreneurs in the social firms or the third sectors. Social entrepreneurs should continuously motivate collaborative social innovation to create social value by connecting creatively between technology and society. Social open innovation by open and creative connections between technology and society should be pursued for sufficient and sustainable social values in the social innovation economy.

The final agencies are governments. They should play the role as the manager of Schumpeterian dynamics. Overly weak subeconomies should be vitalized by governments. Oppositely, overly strong subeconomies should be controlled to motivate sustainable Schumpeterian dynamics. Perhaps the role of governments is not to remedy market failures but to ensure the vitalization of national innovation systems.

Research Question

1. Can we analyze our economy using the Schumpeterian dynamics of open innovation economy system?
2. Can we list vitalizing factors of our Schumpeterian dynamics of OIES?
3. Can we list bottlenecks of our Schumpeterian dynamics of OIES?

References

Belk R (2014) You are what you can access: sharing and collaborative consumption online. J Bus Res 67(8):1595–1600

Benkler Y (2004) Sharing nicely: on shareable goods and the emergence of sharing as a modality of economic production. Yale Law J 114:273–358

Brunswicker S, van de Vrande V (2014) Exploring open innovation in small- and medium-sized enterprises. In: Chesbrough H, Vanhaverbeke W, West J (eds) New frontiers in open innovation. Oxford University Press, Oxford, p 135

Burton-Jones A (2001) Knowledge capitalism: business, work, and learning in the new economy. Oxford University Press, Oxford
Chang H-J (2002) Kicking away the ladder: development strategy in historical perspective. Anthem Press, New York
Chang H-J (2010) 23 things they don't tell you about capitalism. Bloomsbury Publishing, New York
Chesbrough HW (2003) Open innovation: the new imperative for creating and profiting from technology. Harvard Business Press, Boston
Chesbrough HW (2007) Business model innovation: it's not just about technology anymore. Strateg Leadersh 35(6):12–17
Chesbrough HW (2010) Business model innovation: opportunities and barriers. Long Range Plan 43(2):354–363
Chesbrough HW (2013) Open business models: how to thrive in the new innovation landscape. Harvard Business Press, Boston
Chesbrough HW, Di Minin A (2014) Open social innovation. In: Chesbrough H, Vanhaverbeke W, West J (eds) New frontiers in open innovation. Oxford University Press, Oxford, p 223
Christensen JF (2014) Open innovation and industrial dynamics—towards a framework of business convergence. In: Chesbrough H, Vanhaverbeke W, West J (eds) New frontiers in open innovation. Oxford University Press, Oxford, p 94
Christensen JF, Olesen MH, Kjær JS (2005) The industrial dynamics of open innovation—evidence from the transformation of consumer electronics. Res Policy 34(10):1533–1549
Crowther D, Aras G (2008) Corporate social responsibility. E-book. Bookboon. http://bookboon.com/en/defining-corporate-social-responsibility-ebook. (2015.03.14)
Drucker PF (1993) The post-capitalism. Harper and Row Publishers, New York
Drucker PF (1994) Post-capitalist society. Routledge, London
Drucker PF (2014) Innovation and entrepreneurship. Routledge, London
Foray D, Lundvall B (1998) The knowledge-based economy: from the economics of knowledge to the learning economy. In: Need D, Siesfeld GA, Cefola J (eds) The economic impact of knowledge. OECD, Paris, pp 115–121
Ha-Sung J (2014) Korea capitalism: over economy democratizing to righteous economy. Heybooks, Seoul
Hardin G (1968) The tragedy of the commons. Science 162(3859):1243–1248
Holme R, Watts P (1999) Corporate social responsibility. World Business Council for Sustainable Development, Geneva
Honey Bee Network; http://www.sristi.org/hbnew/aboutus.php
Keynes JM (1935) General theory of employment, interest, and money. Atlantic Publishers & Distributors, New Delhi
Kim L (1997) Imitation to innovation: the dynamics of Korea's technological learning. Harvard Business Press, Boston
Kim L (1998) Crisis construction and organizational learning: capability building in catching-up at Hyundai Motor. Organ Sci 9(4):506–521
Kostakis V, Bauwens M (2014) Network society and future scenarios for a collaborative economy. Palgrave MacMillan, Basingstoke
Lee KR (1996) The role of user firms in the innovation of machine tools: the Japanese case. Res Policy 25(4):491–507
Lee J-W (2011) Coupled growth= Win-win growth: game rule without loser. Futur M&B
Lee J-W (2015) Second edition economy of inequality. Humanitas, Seoul
Lee K, Lim C (2001) Technological regimes, catching-up, and leapfrogging: findings from the Korean industries. Res Policy 30(3):459–483
Marx K (1867) Capital, vol 1. Penguin Books, London
Marx K (1978a) Capital, vol. 1: a critique of political economy. Penguin Books, London
Marx K (1978b) Capital, vol 2. Penguin Books, London
Marx K, Simpson R, Ryazanskaya SW (1963) Theories of surplus-value, vol 1. Foreign Languages Publishing House, Moscow

Medio A (1972) Profits and surplus-value: appearance and reality in capitalist production. In: Hunt EK, Schwartz J (eds) A critique of economic theory. Penguin Books, London, pp 312–346

Mortara L, Minshall T (2014) Patterns of implementation of OI in MNCs. In: Chesbrough H, Vanhaverbeke W, West J (eds) New frontiers in open innovation. Oxford University Press, Oxford, p 223

Mulgan G (2006) The process of social innovation. Innovations 1(2):145–162

Ostrom E (1990) Governing the commons: the evolution of institutions for collective action. Cambridge University Press, London

Piketty T (2010) L'économie des inégalités. La Découverte, Paris

Piketty T (2014) Capital in the 21st century. Harvard University Press, London

Piketty T, Goldhammer A (2014) Capital in the twenty-first century. Belknap Press, London

Pisano GP, Verganti R (2008) Which kind of collaboration is right for you? Harv Bus Rev 86(12):78–86

Porter ME, Kramer MR (2011) Creating shared value. Harv Bus Rev 89(1/2):62–77

Rifkin J (2011) The third industrial revolution: how lateral power is transforming energy, the economy, and the world. Macmillan, London

Rifkin J (2014) The zero marginal cost society: the Internet of things, the collaborative commons, and the eclipse of capitalism. Macmillan, London

Rifkin J, Kruger E (1996) The end of work. Social Planning Council of Winnipeg, Winnipeg

Rose C (1986) The comedy of the commons: custom, commerce, and inherently public property. Univ Chicago Law Rev 53(3):711–781

Schumpeter JA (1934) The theory of economic development. Harvard University Press, Cambridge, MA

Schumpeter JA (1939) Business cycles: a theoretical, historical, and statistical analysis of the capitalist process. McGraw-Hill Book Company, New York

Schumpeter JA (1942) Capitalism, socialism, and democracy. Routledge, London

Simon HA (1982) Models of bounded rationality: empirically grounded economic reason, vol 3. The MIT Press, Cambridge, MA

Simon HA (1991) Bounded rationality and organizational learning. Organ Sci 2(1):125–134

Smith A (1937) The wealth of nations. Random House, New York

Sundararajan, A (2013) From Zipcar to the sharing economy. Online article. Harvard Business Review. http://hbr.org/2013/01/from-zipcar-to-the-sharing-eco/(2015.03.14)

Teece DJ, Pisano G, Shuen A (1997) Dynamic capabilities and strategic management. Strateg Manag J 18(7):509–533

Weitzman ML (1985) The simple macroeconomics of profit sharing. Am Econ Rev 75(5):937–953

Wolff EN (1979) The rate of surplus value, the organic composition, and the general rate of profit in the US economy, 1947–67. Am Econ Rev 69(3):329–341

Yun JHJ (2015) How do we conquer the growth limits of capitalism? Schumpeterian dynamics of open innovation economy system. J Open Innov Technol Market Complex 1(17):1–20

Yun JJ, Won D, Hwang B, Kang J, Kim D (2015) Analysing and simulating the effects of open innovation policies: application of the results to Cambodia. Sci Public Policy 42(1):1–18

Zervas G, Proserpio D, Byers J (2014) The rise of the sharing economy: estimating the impact of Airbnb on the hotel industry. Boston University School of Management Research Paper Series, 2013–16

Economic Effects of Open Innovation

3

Abstract

We study the economic effects of open innovation, which has been recognized as an economic phenomenon and economic paradigm that surpasses the management strategies of individual enterprises. First, we look into limited open innovation phenomena such as long tail phenomena and App Store. Next, the economic characteristics of open innovation have been identified: increases of marginal products, economy of diversity, and X-efficiency improvement.

Keywords

Long tail economy • Increases of marginal products • The economy of diversity • X-efficiency

3.1 Introduction

At the supply of smartphones and Web 2.0-based culture and the business practice of communication and sharing have been activated, open innovation has gradually gained recognition as an economic phenomenon rather than a phenomenon at the level of certain individual enterprises. Therefore, the effect of open innovation can be assumed to generate certain economic effects rather than just improve enterprises' productivity.

This chapter is mainly based on the following paper. But, this chapter includes just nearly 65% of the paper. Please read the following full paper:
Yun J.H.J., Cho B.J. (2014) An Exploratory Study of the Economic Effect of Open Innovation. *Journal of Science & Technology Policy Management*. Vol. 5, No. 1

J.J. Yun, *Business Model Design Compass*, Management for Professionals,
DOI 10.1007/978-981-10-4128-0_3

InnoCentive, which is a global open innovation knowledge transaction company, announced that it would develop into a global crowd sourcing platform through a publication on its home page on June 19, 2010. According to Wikipedia, crowd sourcing is opening the processes of production and service so that consumers or the public can participate in order to enhance production efficiency and share profits with participants. This corresponds to user innovation presented by Von Hippel (2006), which describes a type of innovation through users' ideas. InnoCentive regards those who propose problems to be solved as seekers and those who provide solutions to the problems as solvers and seeks both from knowledge consumers in the world. In Korea, typical cases of crowd sourcing have been implemented through a mapmaking project and a group singing project where ordinary citizens participated in singing "Goose's Dream" together (Jeon 2011). Meanwhile, crowd sourcing is defined to be similar to "peer-to-peer producing"—"phenomena in which many people and enterprises openly cooperate to lead innovation and growth."

Prosumers, a compound word of producers and consumers where consumers use products and play the role of producers who participate in the processes of product development and production, have been one of the major players in the twenty-first century knowledge economy (Toffler et al. 1981). That is, the characteristic that consumers act as the main agents in the producers' new product or process innovation is a core element of the twenty-first century knowledge economy. With the activation of the mobile Internet, prosumer phenomena sometimes appear as mash-up phenomena in which information and services provided through the Web are fused together to make new software, services, and databases, among others, or open software phenomena in which program users gradually participate as producers of relevant S/W.

Chris Anderson took notice of the "long-tailed distributions" because the tail part of demand curves is relatively much longer than the head part in the situation where marketing costs or marginal costs for additional sales are very low because of the propagation of the Internet and defined this economic phenomenon as the long-tailed economy. Through analyses of products sold through online stores, such as Netflix, iTunes, Amazon, and Google, he established a law that replaces "The Law of the Vital Few" presented earlier by Vilfredo Pareto indicating that approximately 20% of people possesses 80% of wealth. A Pareto distribution, otherwise known as the Law of the Vital Few, indicates that similar rates to those mentioned above appear in general industries, such as cases where 20% of products account for 80% of all revenues, cases where 20% of the time invested accounts for 80% of the total productivity, or cases where 20% of the time spent by us accounts for 80% of our total productivity (Anderson 2008).

On the other hand, the long-tailed law describes a new form of demand phenomena in which consumers' interest has moved to the remaining 98% of products except for hit products. Although a major effect of long-tailed phenomena is making our preference move from hit products to niche products, if we are more satisfied with what we found, we will consume more products that fall under the long tail (Anderson 2007). The most important source of long-tailed economic phenomena is the arrival of the knowledge-based society in which little costs are incurred in the production and distribution of knowledge and technology. The costs spent in manufacturing and distribution in the traditional economy are close to zero in long-tailed markets dominated by digital products that can be copied and transmitted at almost no cost (Anderson 2007). Therefore, ideas can be considered abundant because they can be infinitely spread in that anybody can freely produce ideas (Anderson 2007). The abundant ideas of users, consumers, related enterprise, and supporting enterprises, among others, become the sources of production and consumption of products that correspond to 98% of niche markets. The new phenomenon in which individuals or small-scaled economy systems are networked in real time through the Internet and make large-scaled fashion and propagation at an immensely high speed through viral effects is sometimes described as nanoeconomics (Jeon 2011). According to Jeon (2011), as the paradigm moves from the existing mass economy of mass production and mass sales to a nano-economy in which certain very trivial consumers emerge as major players, the nano-economy provides service and information that accurately satisfy individual consumers' needs, and the size of transactions between individuals and transactions in small quantities is growing.

The economic characteristics of the world implemented by social media such as Twitter and Facebook are called socialnomics (Qualman 2009). According to Qualman, microblogging contains new potential not only as a system for providing and obtaining ideas but also as a stage of implementation of user-based business models. This is another piece of evidence indicating that consumers and users are coming forward as producers of information and knowledge, joint leaders of innovation, and leaders of open innovation. Socialnomics presents the potential of open innovation even for the production of new products surpassing the limited economic phenomena, such as advertisements and marketing. The unceasing participation and sharing by CrackBerries become the driving force of social economics.[1]

3.2 Phenomena of Open Innovation

3.2.1 Long-Tailed Phenomena

According to a survey conducted by Anderson (2007), in the case of Walmart, the largest record store in the USA as of 2005, the top 200 records in sales account for 90% of record sales. On the other hand, in the case of a record shop named Rhapsodie, a large long tail was formed, and the 25,000th music piece to the 100,000th music piece recorded 25% of the total sales. Furthermore, even songs ranging from the 100,000th to 800,000th position formed a meaningful long-tailed distribution, recording 15% of the total sales. That is, long-tailed phenomena are appearing such that when Rhapsodie increases the number of music pieces they provide, customers scattered throughout the world and buy the songs every month. These long-tailed phenomena also appear in online bookstores, Walmart, Best Buy,

[1] CrackBerry is a newly coined word that combines the words "crack," a sort of narcotics, and "BlackBerry," which refers to users addicted to the use of smartphones, such as BlackBerry.

and numerous other retail stores as well as in all industrial areas, such as the long tail of beer made through the diffusion of small breweries, the long tail of fashion, and the long tail of education made through the growth on online education companies (Anderson 2007, 2008).

The long-tailed economy is promoted through the popularization of production tools, the popularization of distribution structures, and the connection between demand and supply, among others (Anderson 2007). The popularization of production tools converts passive producers into active ones, thereby acting as a major drive to create the economy of diversity. For instance, the phenomenon that readers exert influence through online book reviews, among others, clearly shows the view of changes of passive customers into active customers. Through these processes, the participatory producer-oriented economy of diversity appears from consumer-oriented sales. In other words, popularized production tools that enable customers to buy products more easily at cheaper prices increase the number of producers enormously to realize the economy of diversity.

The reduction of distribution costs achieved through mobile data traffic, among others, effectively increases the liquidity of markets located at the tail part by enabling everybody to provide content to many people without any additional cost or with little cost. This reverses the marginal product diminishing phenomena occurring because of distribution cost increases and becomes a major source to cause increased marginal product phenomena.

3.2.2 App Store Phenomena

The App Store is an online sales service operated by the IT enterprise Apple. It deals with products (application) that can be used on iPhones as items and has a global online open market that provides business models that can be used by anybody in the next generation, that is, those who use the Internet, specifically, the mobile Internet. This App Store has products that are not sold in large quantities but are still sold continuously, occupying a part of the long tail, and thus provides business models suitable for individuals or small and medium enterprises rather than large enterprises.

The App Store played the role of a window to open ways for individual developers or development firms to develop applications operating in the iPhone OS, using the software development tools provided by Apple, and sell the applications firsthand. Apple eliminated interference by mobile carriers and specified that developers should take 70% of profits from sales to motivate them to develop products. After opening in July 2008, the App Store had 65,000 registered applications in 1 year, and the number exceeded 100,000 in November 2009. The App Store recorded 1.5 billion downloads in July 2009 and made a new record in September 2009 when it exceeded two billion downloads.

In addition to the App Store on Apple's iPhone, other App Store models for smartphones, such as Google's Android Market and Nokia's Ovi Store, are currently being expanded gradually to user participation or user innovation-based open innovation business models not only for the Internet TV of Google, Samsung, and Apple but also for diverse other businesses, including automobiles. That is, the concrete form of the App Store business is gradually expanding into new business models (Amit and Zott 2012; Holzer and Ondrus 2009; Jansen et al. 2009; Müller et al. 2011; Osterwalder and Pigneur 2010; Yun and Mohan 2012).

In App Stores, user-based open innovation is activated through general users' active participation (Yun 2010). This active participation in production activities by users starts to create the economy of diversity. That is, the production and supply of quite diverse applications become possible. This is also the largest source of the current competitiveness of Apple App Store and iPhone.

The global marketing and sales networks possessed by the App Store enable product production and sales even with little expenditures of marketing costs or marginal costs. This becomes a direct cause that brings about marginal productivity increasing. That is, the situation where anybody can sell new products through App Store global networks without any separate marketing or distribution costs creates phenomena of increasing marginal products, reversing the trend of diminishing marginal products resulting from increasing marginal costs. Finally, the core of the App Store phenomena is that the consumers of applications are the base of development of App Store applications (Yun 2010).

3.3 Economic Effects of Open Innovation

3.3.1 Marginal Product Increasing

The most important characteristics brought about by the activation of smartphone-based mobile data traffic in the knowledge-based economy include, first, phenomena of increasing marginal products.

When mobile data traffic is activated in a knowledge-based economy, the inflows of new knowledge and technologies from the outside rapidly increase as open innovation of knowledge and technologies is activated, as labor inputs increase, or as the amount of capital increases. Furthermore, more effective marketing or technology innovation occurs through the outflows of unutilized knowledge and technologies to the outside. This is a new characteristic of the knowledge-based economy—the mobile data traffic open innovation economy.

Along with the acceleration of a knowledge-based economy in modern society, the phenomenon of increasing marginal products appears as a reduction of business fluctuations in the world economy and the acceleration or continuation of economic growth in countries where mobile data traffic open innovation has developed. According to Romer (1990), technological advances are the driving force of economic growth, and the outcomes of technological advances are created by research and development. According to Romer, the core of technological advances is the creation of new knowledge. When new knowledge has been created through research and development, the knowledge is widely utilized by society members, in particular, by enterprises, as part of open innovation and the productivity of production factors, such as labor and capital, is enhanced. This is because of the absence of rivalry of knowledge and technology as semipublic goods (Rivera-Batiz and Romer 1991; Romer 1986, 1990).

3.3.2 Economy of Diversity

The development of a knowledge-based society and the activation of open innovation have characteristics of product diversification and the growth of niche markets. The unceasing open innovation by users or consumers and related enterprises and supporting enterprises enables continuous appearance of new niche innovation products. These products are delivered to consumers through smartphone mobile data traffic without any additional marketing costs, and the size of niche markets continuously grows.

This economy of diversity is gradually being accelerated as the opportunity and possibility of participation in the production of niche innovation products through users' participation in user-based open innovation are expanded based on the existing Internet and mobile data traffic (Di Gangi and Wasko 2009; Franke and Piller 2004; Henkel and Von Hippel 2005; Luthje et al. 2003; Morrison et al. 2000; Perez and De Pablos 2003; Sundbo and Toivonen 2011; Von Hippel 2009). The activation of the economy of diversity means the erosion of the economy of scale. In reality if open innovation is activated, the economy of scale intended to reduce costs is gradually declining, and the economy of scope is being activated. The concept of the long-tailed economy describes the shape of the graph but does not describe the essence of certain economic phenomena. The economy of diversity can be defined as the activation of niche markets and the expansion of the diversity of niche products from commons, customers, and users, which are the essence of the long-tailed economy (Dietz et al. 2003; Ostrom 1990, 2009).

3.3.3 X-Efficiency Enhancement

The activation of open innovation enables the renewal of the maximum value of products that can be made by enterprises by investing certain labor and capital. That is, no allocative efficiency can be maximized by new technology-based new products or new process innovation. Therefore, the activation of open innovation boils down to the enhancement of X-efficiency.

Even in cases where enterprises' X-inefficiency appears because of monopoly, the activation of innovation utilizing users' or consumers' new innovation ideas or user-based open innovation by users or consumers can enhance the X-efficiency of the relevant product production to offset the X-inefficiency caused by monopoly. For instance, the continuous or gradual expansion of market sizes and operating profits appearing in Amazon.com, Apple App Store, iTunes, etc., despite their monopolistic nature, is arguably evidence of the X-efficiency of open innovation that offsets the X-inefficiency of monopoly (De Alessi 1983; Fu and Heffernan 2007; Leibenstein 1978).

Among the conditions for open innovation activities intended to enhance X-efficiency, three conditions have been presented from the viewpoint of the success factors of user-based open innovation in App Store (Yun 2010). These conditions are production tool kits with high user convenience and completeness, transparent and immediate guarantee of profits of participants in S/W production, and clear attribution of intellectual properties from open innovation to users. In other words, if transparent and clear systematic and economic devices are installed, which will enable the users or consumers of the relevant product to participate in product innovation and obtain guaranteed profits, user-based open innovation will be immediately activated, and X-efficiency will be enhanced.

Research Question

1. Identify and explain the economic effect of open innovation by evidence in your life.
2. Examine and explain the open innovation of App Stores such as Apple App Store or Android Market.
3. Find long tail examples in your life and analyze and explain them by the open innovation concept.

References

Amit R, Zott C (2012) Creating value through business model innovation. MIT Sloan Manag Rev 53(3):41–49

Anderson C (2007) The long tail: how endless choice is creating unlimited demand. Random House, London

Anderson C (2008) The long tail: why the future of business is selling less of more. Hyperion Books, New York

De Alessi L (1983) Property rights, transaction costs, and X-efficiency: an essay in economic theory. Am Econ Rev 73(1):64–81

Di Gangi PM, Wasko M (2009) Steal my idea! Organizational adoption of user innovations from a user innovation community: a case study of Dell IdeaStorm. Decis Support Syst 48(1):303–312

Dietz T, Ostrom E, Stern PC (2003) The struggle to govern the commons. Science 302(5652):1907–1912

Franke N, Piller F (2004) Value creation by toolkits for user innovation and design: the case of the watch market. J Prod Innov Manag 21(6):401–415

Fu XM, Heffernan S (2007) Cost X-efficiency in China's banking sector. China Econ Rev 18(1):35–53

Henkel J, Von Hippe E (2005) Welfare implications of user innovation. Essays in Honor of Edwin Mansfield. Springer US, pp 45–59

Holzer A, Ondrus, J (2009) Trends in mobile application development. International conference on mobile wireless middleware, operating systems, and applications, Springer, Berlin, Heidelberg

Jansen S, Finkelstein A, Brinkkemper S (2009) A sense of community: a research agenda for software ecosystems. Paper presented at the ICSE 2009 31st international conference on software engineering: companion volume.

Jeon B (2011) A perspective of domestic appstores compared with global appstores. In: Kim T, Adeli H, Robles RJ, Balitanas M (eds) Advanced communication and networking. Communications in computer and information science, vol 199. Springer, Berlin, pp 271–277

Leibenstein, H. (1978). General X-efficiency theory and economic development, New York, Oxford University Press.

Luthje C, Herstatt C, Von Hippel E (2003) The dominant role of "local" information in user innovation: the case of mountain biking. TUHH, Hamburg

Morrison PD, Roberts JH, Von Hipple E (2000) Determinants of user innovation and innovation sharing in a local market. Manag Sci 46(12):1513–1527

Müller RM, Kijl B, Martens JK (2011) A comparison of inter-organizational business models of mobile App stores: there is more than open vs closed. J Theor Appl Electon Commer Res 6(2):63–76

Osterwalder A, Pigneur Y (2010) Business model generation: a handbook for fisionaries, game changers, and challengers. Wiley.com.

Ostrom E (1990) Governing the commons: the evolution of institutions for collective action. Cambridge University Press, Cambridge

Ostrom E (2009) Understanding institutional diversity. Princeton University Press, Princeton

Perez JR, De Pablos PO (2003) Knowledge management and organizational competitiveness: a framework for human capital analysis. J Knowl Manag 7(3):82–91

Qualman E (2009) Socialnomics. Hoboken, New Jersey

Rivera-Batiz LA, Romer PM (1991) Economic integration and endogenous growth. Q J Econ 106(2):531–555

Romer PM (1990) Endogenous technological change. J Polit Econ 98:S71–S102

Sundbo J, Toivonen M (2011) User-based innovation in services. Edward Elgar Publishing, Northampton

Toffler A, Longul W, Forbes H (1981) The third wave. Bantam Books, New York

Von Hippel (2006) *Democratizing Innovation*. The MIT Press. Cambridge, Massachusetts

Von Hippel E (2009) Democratizing innovation: the evolving phenomenon of user innovation. Int J Innov Sci 1(1):29–40

Yun J-H (2010) Open innovation business foundation policy management theory. Book & World, Seoul

Yun JHJ, Mohan AV (2012) Exploring open innovation approaches adopted by small and medium firms in emerging/growth industries: case studies from Daegu–Gyeongbuk region of South Korea. J Sci Technol Policy Manage 12(1):1–19

Open Innovation Policy in National Innovation System

4

Abstract

The use of purposive inflows and outflows of knowledge to accelerate internal innovation and to expand the markets by external use of innovation has recently become essential not only in business strategies but also national policies. According to a literature review, open innovation policies should have three aspects, namely, knowledge and technology production, distribution, and consumption. If different levels of open innovation policies are introduced into the national innovation system (NIS), different effects occur not only in the short term but also the long term. The fundamental goal of open innovation policy should be vitalizing Schumpeterian dynamics of the open innovation economy system.

Keywords

Open innovation policy • National innovation system

This chapter is mainly based on the following papers. More than 40% of following papers were changed at this chapter. Please read the full papers.

Yun J.H.J, Won D.K., Hwang B.Y., Kang J.W., Kim D.H. (2015). Analysing and simulating the effects of open innovation policies: Application of the results to Cambodia. *Oxford Journal of Science and Public Policy*, 42, pp. 743–760.

Yun J.H.J., Jung W.Y. (2013). Open Innovation Policies in Asian Countries. *Asian Research Policy*. Vol. 4, No. 1.

J.J. Yun, *Business Model Design Compass*, Management for Professionals,
DOI 10.1007/978-981-10-4128-0_4

4.1 Introduction

The concept of the national innovation system (NIS) was mainly developed by three scholars: Freeman, Lundvall, and Nelson (Fagerberg and Sapprasert 2011). Common features can be found from the discussions in the process of the settlement of the concept of the national innovation system. Freeman (1987) took note of the network of institutions and new technologies made by the network. Lundvall (1992) emphasized interactions among elements for the production, diffusion, and use of economically valuable knowledge. Nelson (1992) emphasized interactions among organizations that enhance the records of innovation of enterprises in the country and the OECD (1997) focused on the distribution of technology and information (Patel and Pavitt 1994). Overall, the creation of innovation performance through the production, distribution, and consumption of knowledge and technology among parties in the country was generally defined as the national innovation system (Metcalfe 1995).

National innovation systems have components of quite diverse natures depending on the subjects to be analyzed, such as enterprises, state organs, and research institutes. Therefore, national innovation systems that have diverse components can be designed depending on the study purposes and content of the research. However, national innovation systems related to production, distribution, and consumption of knowledge should be designed. The diverse relationships among the elements of national innovation systems basically have the characteristics of network relations. National innovation system components have characteristics of networks that are far from single directions or hierarchical structures. In addition, several studies analyzed changes in the time series of national innovation systems, such as Wong (2011), Hung and Whittington (2011), and Metcalfe and Ramlogan (2008).

4.2 Open Innovation Policy

Recently, the concept of open innovation is expanding from a micro firm level to a macro national innovation system (De Jong et al. 2008; Fu and Xiong 2011; Herstad et al. 2010; Lee et al. 2012; Santonen et al. 2007; Wang et al. 2012).

First, diverse policies for increasing open innovation of firms are introduced through networking, collaboration, corporate entrepreneurship, intellectual property management, and research and development (De Jong et al. 2008). Insofar as open innovation is about "open" business models for innovation, countries' framework conditions (i.e., product and labor markets, IPR and competition policies, a strong public research base, etc.) are extremely important policy levers (De Backer 2008). Based on evidence from firm-level case studies, the evolution of policies and practices concerning open innovation in China was analyzed in the same context (Fu and Xiong 2011).

Open innovation processes occurring on an international scale are also macro-level category of open innovation (Patra and Krishna 2015). In this category, public policy now needs to carefully balance between the following: (a) promoting the formation of international linkages for knowledge sourcing and information exposure, (b) providing incentives for domestic industry intramural R&D for building absorptive capacity and knowledge accumulation, and (c) sustaining domestic networking to allow accumulated knowledge to diffuse and recombine (Herstad et al. 2010). This category is different from foreign direct investment or trade in that global open innovation focused on distribution of knowledge internationally even though FDI and trade also (Helpman 2006) have technology transfer effects (Helpman et al. 2003; Kokko et al. 1996).

Third, the open innovation in the public sector is another macro-level open innovation in that it deals with open innovation beyond the firm's boundary (Yun 2015). For example, the USA, Australia, and Singapore developed open innovation policies at the national level, facilitating a positive innovative climate (Lee et al. 2012).

Fourth, a special national open innovation system is proposed to support any national innovation system. An example is the general innovation triangle framework that consolidates the national open innovation system (NOIS) for supporting the Finnish national system of innovation (Santonen et al. 2007).

Chesbrough (2003) models the relationship between national innovation system and OI in his illustration of how structural changes in the US national innovation system have created a widely distributed knowledge landscape since World War II because of the greater availability of highly educated workers, venture capital, and state-of-the-art knowledge (Wang et al. 2012). In fact, the relationships between OI and national innovation systems have quite important effects on policies. Few studies have analyzed these two together because, whereas the former focuses on transfers of knowledge and technology across the boundaries of enterprises as individual firms, the latter focuses on the production, distribution, and consumption of knowledge and technology in each country. Because OI activation at the level of enterprises is an important drive in the enhancement of the productivity of national innovation systems, diverse discussions on the necessity and methods of the development of policies that can activate OI have been raised (Jong et al. 2010). Not only OI activation policies for fixing market failure but also more active roles of the government for enhancing open innovation at the level of the systems of innovation are required. If enterprises open up their innovation process, they will obtain better results (Chesbrough 2003). Similarly, innovation is the result of complex and intensive interactions among various actors, according to Lundvall (1992). That is, many similarities between open innovation and systems of innovation models exist (Jong et al. 2010).

Few policy studies to promote OI have been carried out because policy studies in this area began only recently. Policies for enlarging the OI of target national innovation systems are diverse, according to the researchers or research goals. However, there are some common OI policies among the three research groups. The characteristics and essence of OI policies may be established in terms of drawing policies commonly presented in recent studies of OI policies. These are concurrent among knowledge production, distribution, and consumption. Policy for open innovation in Table 4.2 does not have intention to motivate open innovation internationally directly.

4.3 National Innovation System Dynamic Model of Open Innovation Policy

4.3.1 Model Building

Based on almost all national innovation system studies, a national innovation system causal loop diagram was established. This was prepared on the basis of the conceptual and logical positive feedback loop of national innovation systems, that is, systematic links that connect knowledge accumulation and activation of inflows of global knowledge as well as increase in technical innovation. In addition, enhancement of product and process innovation is connected with the improvement of national competitiveness through increased output. In the end, increases in new research and development investments trigger a new loop.

As shown in Table 4.1, policies for promoting open innovation are aimed at enhancing the value of knowledge first. The enhanced value of knowledge promotes research and development investments in the country and inflows of knowledge and technology from foreign countries in diverse forms and through diverse channels. In the stage of intervention by OI policies for activating OI, increases in the value of knowledge are triggered by the activation of OI, and the effects of such increases on domestic and foreign elements are explicitly reflected in the national innovation system causal loop diagram.

Also, dissipation of the price advantage or differentiation advantage by the enterprises that own technologies at the enterprise level is indicated as a negative feedback loop (Porter 1985). Although OI promotes knowledge production and distribution and brings about enhancement of national competitiveness at the national innovation system level, the activation of OI in relation to certain technologies dissipates the competitive advantage of the relevant companies (Baden-Fuller and Stopford 1992). Thus, the effects of OI policies on national innovation systems can be simulated more practically by including a negative feedback loop of OI.

The relation among individual variables constituting a causal loop has been sufficiently elucidated already by many previous studies, as shown in Table 4.1. However, this study shows systematic and creative composition of the causal relationships among related variables as a causal loop diagram.

Open innovation has positive and negative effects together. At a microlevel, open innovation has a reverse U-curve effect on a firm's performance because of its negative effects (Laursen and Salter 2006). At a macro-level, like the Latin American cases, too severe open innovation can have a negative effect on the national innovation system. All these negative effects appear in the negative feedback loop in Fig. 4.1 (AMERICA 1950; Arocena and Sutz 2000).

Table 4.1 Grounds for setting the relationships between variables in the causal loop

Relationships between variables	References
Open innovation vitalization → + increasing value of knowledge	Chesbrough et al. (2011) Jong et al. (2010) Wang et al. (2012)
Open innovation vitalization → – technological competitive advantage	Porter (1985) Baden-Fuller and Stopford (1992)
Increasing value of knowledge → + national R&D vitalization → + R&D investment	Richardson (1997) Sterman (2000) Romer (1990) Ford and Sterman (1998)
Increasing value of knowledge → + global knowledge influx → + technical innovation	Castellacci and Natera (2012) Wong (2011) Niosi (2010) Sterman (1985)
R&D investment → + accumulating knowledge → + technical innovation → + new product, new process → + output increase → + national competitiveness → + R&D investment	Freeman (1987) Lundvall (1992) Nelson (1992) OECD (1997) Metcalfe and Ramlogan (2008) Samara et al. (2012) Kim and An (2011) Won and Yun (2005)

Source: Yun et al. (2015)

4.3.2 Different Levels of Open Innovation Policy

In this study, as shown in Fig. 4.2, OI polices were divided into low-level OI policies, medium-level OI policies, which cover knowledge production and distribution, and high-level OI policies, which cover knowledge production, distribution, and consumption. The OI indexes of these policies were set as 0.1, 0.5, and 0.8 to simulate the results.

The level at which a government's OI policies are not properly exercised and at which enterprises' OI practices in the market are insignificant was assigned an open innovation index of 0.1. The level at which the government's OI policies are partially exercised, that is, concentrated mainly on the production and distribution of knowledge and technology using some diverse measures of OI policies and OI practices in the market are activated moderately, was assigned an open innovation index of 0.5. The final level at which the government's OI policies are fully exercised, that is, exercised using diverse OI policy measures in all the areas of the production, distribution, and consumption of knowledge and technology and OI practices in the market are activated, was assigned an open innovation index of 0.8.

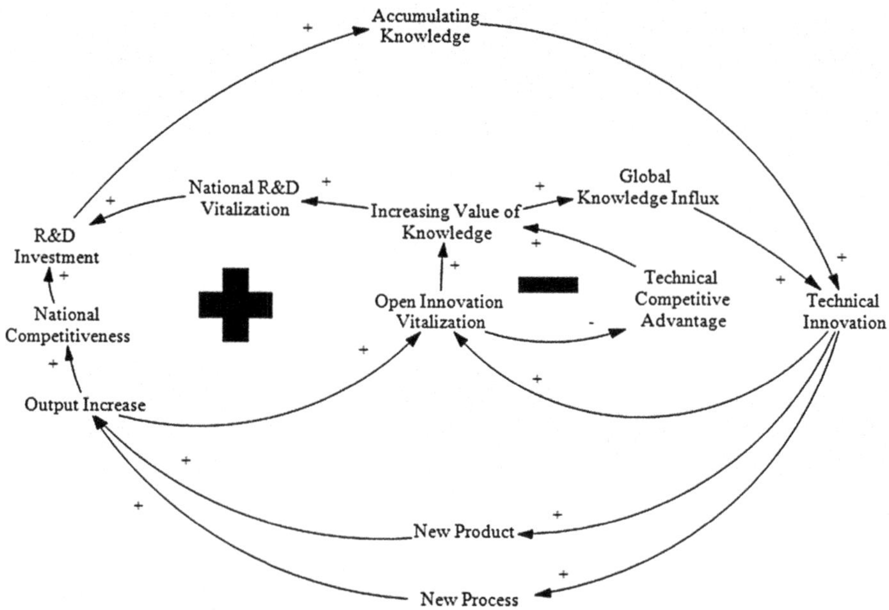

Fig. 4.1 National innovation system causal loop to analyze the effects of OI policy on national innovation system (Source: Yun et al. (2015))

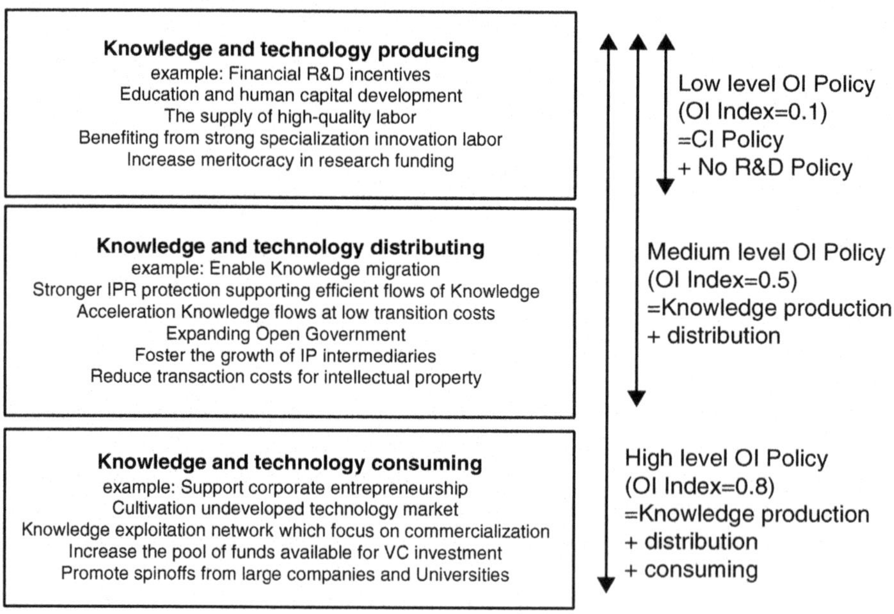

Fig. 4.2 OI policy as knowledge production, distribution, and consumption (Source: Yun et al. (2015))

4.4 Scope of Open Innovation Policy

Open innovation policies are intended to promote production, distribution, and consumption of knowledge and technology across the boundaries of enterprises in the country and have different characteristics from those of existing industrial policies, scientific technology policies, and research and development policies. First, let us examine the differences between industrial policies and open innovation policies. Industrial policies are diverse policies intended to promote the development of certain industries, which range from conventional industrial policies that correct market failure related to certain policies to activate the relevant industries to Schumpeterian industrial innovation policies that activate the innovation systems of the relevant industries to lead to enterprises' product innovation or process innovation. Open innovation policies are different from existing industrial policies in that they do not predefine industries or products per se and that they promote the production and distribution of protected knowledge as well as protectable knowledge instead of a certain sector.

Second, scientific technology policies focus on the activation of science bases in the country and the production of economically valuable technologies. Therefore, for the production of scientific and source technologies mainly by universities and the production of applied and developed technologies by government-funded research institutes or enterprises not only the Ministry of Education, Science and Technology of Korea but also almost all governmental departments become main agents of scientific technology policies, including individual departments in certain areas and the Ministry of Knowledge Economy focusing on applied and developed technologies. However, open innovation policies consider the distribution and consumption of knowledge and technology as importantly as the production of knowledge.

Third, research and development policies refer to various kinds of policies related to research and development investments necessary for the production of basic scientific and source technologies as well as various kinds of necessary applied technologies. That is, research and development policies focus on primary production of knowledge.

On the other hand, since open innovation policies aim at the promotion of the production, distribution, and consumption of knowledge across the boundaries of enterprises, encouraging enterprises to be equipped with the research and development capacities necessary for them to actively serve those functions as knowledge brokers in their areas should be included as important subjects of the policies.

Industrial innovation processes are becoming more open. The large, vertically integrated R&D laboratory system of the twentieth century is giving way to more vertically disintegrated networks of innovation that connect numerous companies into ecosystems (Chesbrough and Vanhaverbeke 2011). Based on the above discussion, the categories of open innovation policies are concretely presented in Table 4.2.

Table 4.2 Concrete contents of open innovation policies

Concrete contents of open innovation presented by Chesbrough and Vanhaverbeke	Nature of the relevant policies, knowledge production, distribution, and consuming
1. Education and human capital development – increase meritocracy in research funding in the boundary – support the mobility of researchers among universities, national laboratories, and companies	– production + distribution – distribution
2. Financing open innovation: the funding chain – increase the pool of funds available for VC investment – support the formation of university spin-offs to commercialize research discoveries	– consuming – distribution + consuming
3. Adopt a balanced approach to intellectual property – reduce transaction costs for intellectual property – foster the growth of IP intermediaries – rebalance university IP policies so broad diffusion of publicly funded research results is easier rather than focusing on royalty income alone	– distribution – distribution + consuming – producing
4. Promote cooperation and competition – shift support from national champions toward SMEs and start-up companies – promote spin-offs from large companies and universities – focus on innovation networks	– distribution +consuming – consuming
5. Expand open government – accelerate the publication of government data – use open innovation processes in government procurement. – support private commercialization of government-funded technology	– producing + distribution – producing + distribution – distribution + consuming

Source: Yun and Jung (2013)

4.5 Issues of Open Innovation Policies

4.5.1 Open Innovation as a Democratizing Innovation

The innovation proposed by Schumpeter was innovation based on creative entrepreneurs' entrepreneurship at the beginning and the discussion developed into innovation by groups of large enterprises later (Schumpeter 1942). Thereafter, diverse innovation discussions were developed in evolutionary economics and innovation studies, and most of the discussions were centered on the theory of innovation systems. The theory of innovation systems is clearly more progressive that not only the neoclassical theory but also the Keynesian theory in that all of discussions of government's policies' intervention in system failure, national innovation systems, and regional innovation systems have a theoretical basis of the government's more active intervention in markets. However, policy in this area in most Asian countries including not only Korea but also Japan and China has characteristics distinguished from innovation studies in Europe or in the USA that are linked to political progressiveness in that they approach innovation policies from practical viewpoints. Therefore, the fact that the open innovation has an aspect of democratizing innovation as strong as user innovation should not be overlooked in that the open innovation becomes the basis of the innovation logics of SMEs or individual business founders based on creative ideas. That is, rather than having a value as an innovation strategy of large enterprises having sufficient research and development capabilities or a policy for the strategy, open innovation can be a strategy more suitable for SMEs that promote creative innovation and sustainable growth based on diverse external ideas. In addition, its value as a means of the start-up of individuals' open innovation business models based on ideas and knowledge existing in the world as a means of sustainable economic and social growth should not be overlooked (Yun 2010).

4.5.2 Open Innovation Is "Very Unique"

Diverse discussions treat that the theory of open innovation is not unique. They point that the theory of open innovation can already be found in existing economic or business administration theories such as the discussion on the effect of external economy that surpasses existing economic effects in economic theories or customer relationship management that took notice of customers' demands, expectations, or opinions in business administration. However, the importance and value of the discussion on open innovation that explicitly took notice of enterprises' or diverse economic units' pursuit of new innovation based on external knowledge and technology that they did not create or their transfer to the outside or utilization of internal knowledge and technology that are not utilized cannot be overlooked. In particular, in Asian countries where the importance of a technology-based economy has been just established in economic systems, cultures that make enterprises and economic units be unwilling to acknowledge the value of open utilization of technologies are overflowing. It is worth observing whether the culture in Japan, which is hostile to open innovation, is another hidden reason the Japanese economy, which had been even expected to go beyond the economy of the USA, fell into recession at the time when knowledge distribution and consumption became more important than internal production of knowledge. Open innovation is a phenomenon having quite unique characteristics that enable enterprises or governments to obtain high economic profits only when they have invested considerable amounts of finances and strategic costs.

4.5.3 Open Innovation Needs Enough R&D Investment

The paradigm of open innovation has appeared following the advent of knowledge-based economy in which the amounts of knowledge and technology existing outside innovating bodies, such as enterprises, have become sufficiently large. In addition, in particular, open innovation has been watched as an innovation strategy when enterprises that have accumulated sufficient technology and knowledge on their own started to take notice of innovative external technologies or ideas not owned by them. In other words, only those enterprises and countries that have sufficient innovation capabilities can be equipped with the insight and ability to realize the necessity of external knowledge and technologies and acquire the knowledge and technologies. This is the reason that the current core technology management strategies of global leading enterprises in the USA and in Europe are open innovation strategies. The reason Samsung recently established a large research institute in Silicon Valley and had the research institute take full charge of open innovation strategies is also based on this context. The reason technology-based global cutting-edge enterprises and countries where world class cutting-edge industries have developed are leading in open innovation strategies and policies is that only these enterprises and countries are equipped with sufficient internal innovation capabilities based on adequate research and development investments, and they pursued open innovation capabilities.

In other words, open innovation strategies and policies are not something for which internal research and development investments are unnecessary but are strategies and policies that must be pursued on the basis of sufficient internal research and development investments. Asian enterprises or countries that are just being equipped with their own research and development capabilities should never overlook the necessity of internal research and development investments as they face the open innovation strategies and policies pursued by Western advanced countries and cutting-edge enterprises.

Open innovation policies are characterized by the fact that they focus on knowledge distribution and consumption in order to promote human and technology transfer and business start-ups based on the relevant technologies.

4.6 Conclusion

What should be the goal of open innovation policy to conquer the growth limits of capitalism?

First, the fundamental goal should be vitalizing Schumpeterian dynamics of the open innovation economic system. Increasing the dynamics will equate with the development of economy output in quantity or quality.

Second, the open innovation subeconomy system that is based on SMEs and start-ups should be increased by motivating technology or knowledge distribution and an open innovation relationship with big businesses.

Third, the closed innovation subeconomy system that is based on big businesses should be increased by motivating technology or knowledge creation and an open innovation relationship with SMEs and social firms.

Fourth, the social innovation subeconomy system that is based on social entrepreneurs' collaboration innovation should be increased by motivating technology or knowledge consumption and an open innovation relationship with big business and SMEs.

Research Question

1. Develop your own open innovation policy that can motivate open innovation at your national innovation system.
2. Examine and identify policies that have been disturbing open innovation at your national innovation system.

References

AMERICA OL (1950) The economic development of Latin America and its principal problems

Arocena R, Sutz J (2000) Looking at national systems of innovation from the south. Ind Innov 7(1):55 75

Baden-Fuller C, Stopford J (1992) Rejuvenating the mature business. Routledge, New York

Castellacci F, Natera JM (2012) The dynamics of national innovation systems: a panel cointegration analysis of the covolution between innovative capability and absorptive capacity. Res Policy 42(3):579–594

Chesbrough H (2003) Open innovation: the new imperative for creating and profiting from technology. Harvard Business School Publishing Corporation, Boston

Chesbrough H, Vanhaverbeke W (2011) Open innovation and public policy in Europe. ESADE Business School, Science Business Innovation Board, Barcelona

Chesbrough H et al (2011) Open innovation and public policy in Europe. ESADE Business School, Science Business Innovation Board

De Backer K (2008) Open innovation in global networks. Organization for Economic Co-Operation and Development, Paris

De Jong JP, Vanhaverbeke W, Kalvet T, Chesbrough H (2008) Policies for open innovation: theory, framework and cases. Zoetermeer, Tarmo Kalvet

Fagerberg J, Sapprasert K (2011) National innovation systems: the emergence of a new approach. Sci Public Policy 38(9):669–679

Ford D, Sterman J (1998) Dynamic modeling of product development processes. Syst Dyn Rev 14(1):31–68

Freeman C (1987) Technology and economic performance: lessons from Japan. Pinter, London

Fu X, Xiong H (2011) Open innovation in China: policies and practices. J Sci Technol Policy China 2(3):196–218

Helpman E (2006) Trade, FDI, and the organization of firms. National Bureau of Economic Research, Cambridge, MA

Helpman E, Melitz MJ, Yeaple SR (2003) Export versus FDI. National Bureau of Economic Research, Cambridge, MA

Herstad SJ, Bloch C, Ebersberger B, Van De Velde E (2010) National innovation policy and global open innovation: exploring balances, tradeoffs and complementarities. Sci Public Policy 37(2):113–124

Hung S-C, Whittington R (2011) Agency in national innovation systems: institutional entrepreneurship and the professionalization of Taiwanese IT. Res Policy 40(4):526–538

Jong J, Kalvet T, Vanhaverbeke W (2010) Exploring a theoretical framework to structure the public policy implications of open innovation. Technol Anal Strat Manage 22(8):877–896

Kim DH, An SG (2011) System dynamics modeling for the allocation of national R&D investment. Korean Sys Dyn Res 12:153–176

Kokko A, Tansini R, Zejan MC (1996) Local technological capability and productivity spillovers from FDI in the Uruguayan manufacturing sector. J Dev Stud 32(4):602–611

Lee SM, Hwang T, Choi D (2012) Open innovation in the public sector of leading countries. Manag Decis 50(1):147–162

Lundvall B-Å (1992) National innovation systems: towards a theory of innovation and interactive learning. Pinter, London

Metcalfe S (1995) The economic foundations of technology policy: equilibrium and evolutionary perspectives. In: Stoneman P (ed) Handbook of the economics of innovation and technological change. Blackwell Publishers, Oxford

Metcalfe S, Ramlogan R (2008) Innovation systems and the competitive process in developing economies. Q (Marxt & Brunner) Rev Econ Finance 48(2):433–446

Nelson RR (1992) National innovation systems: a retrospective on a study. Ind Corp Change 1(2):347–374

Niosi J (2010) Building national and regional innovation systems: institutions for economic development. Edward Elgar, Northampton

Organisation for Economic Co-operation and Development (OECD) (1997) National innovation systems. OECD, Paris

Patel P, Pavitt K (1994) The nature and economic importance of national innovation systems. STI Rev OECD Paris 14:9–32

Patra SK, Krishna VV (2015) Globalization of R&D and open innovation: linkages of foreign R&D centers in India. J Open Innov Technol Market and Complex 1(1):1–24

Porter M (1985) Competitive advantage. Free Press, New York

Richardson G (1997) Problems in causal loop diagrams revisited. Syst Dyn Rev 13(3):247–252

Romer P (1990) Endogenous technological change. J Polit Econ 98(5):S71–S102

Samara E, Georgiadis P, Bakouros I (2012) The impact of innovation policies on the performance of national innovation systems: a system dynamics analysis. Technovation 32(11):624–638

Santonen T, Kaivo-oja J, Suomala J (2007) Introduction to national open innovation system (NOIS) paradigm. A preliminary concept for interchange. FFRC eBooks, 8, 2007

Schumpeter J (1942) Capitalism, socialism and Demodcracy. Harper, New York

Sterman J (1985) The growth of knowledge: testing a theory of scientific revolutions with a formal model. Technol Forecast Soc Change 28(2):93–122

Sterman J (2000) Business dynamics: systems thinking and modeling for a complex world. Irwin McGraw-Hill, New York

Wang Y, Vanhaverbeke W, Roijakkers N (2012) Exploring the impact of open innovation on national systems of innovation: a theoretical analysis. Technol Forecast Soc Change 79(3):419–428

Won D-K, Yun JJ (2005) The new Korea innovation systems: an application of system dynamics approach (in Korean). Kor Syst Dyn Res 6(1):71–98

Wong C-Y (2011) Rent-seeking, industrial policies and national innovation systems in Southeast Asian economies. Technol Soc 33(3–4):231–243

Yun J (2010) Open innovation venturing and policy management. KeyoungMun Press, Seoul. (in Korean)

Yun JJ (2015) How do we conquer the growth limits of capitalism? Schumpeterian dynamics of open innovation. J Open Innov Technol Market Complex 1(1):1–20

Yun JHJ, Jung WY (2013) Open innovation policies in Asian countries. Asian Res Policy 4(1):52–66

Yun JHJ, Won DK, Hwang BY, Kang JW, Kim DH (2015) Analysing and simulating the effects of open innovation policies: application of the results to Cambodia. Oxford J Sci Public Policy 42:743–760

Part III

Open Innovation Strategy of Firm

Concept, Structures, and Decision Factors of Open Innovation

5

Abstract

This paper first explains and demonstrates the emergence of open innovation. We also show the structure of open innovation among manufacturing firms and service firms. Last, we show the four factors of open innovation of small- and medium-sized enterprises (SMEs). Third, we look into the open innovation of small and medium enterprises. Fourth, we examine the difference in open innovation according to the product life cycle. Fifth, we analyze the role of internal open innovation attitude regarding external open innovation. Sixth, we talk about the difference in open innovation according to whether a firm is modular or not. Seventh and last, we discuss the difference in open innovation according to regional innovation systems or clusters.

This chapter is mainly based on the following papers:

Yun J.H.J. Ryu G.W. (2009) A Preliminary Study On the Theory and Phenomena of Korea Open Innovation. *International Commerce and Information Review. Vol. 11, No.* 2(in Korean)

Yun J.H.J., Jung W.Y. (2013) Open Innovation of SMEs in Manufacturing from OI Structure Model, *Asia Pacific Journal of Innovation and Entrepreneurship*. Vol. 7, No. 1,

Yun J.H.J., Kim B.T. (2011). A preliminary study on New Innovation Phenomena based on IT Technology Development, *Journal of Daegu Gyeongbuk Development Institute*. Vol. 10, No. 1 (in Korean)

Yun J.H.J., Ryu G.W., Jung W.Y. (2013) Is There Any Difference in the Effect of Open Innovation According to the Product Life Cycle?: Revisiting the Relationship Between Product Life Cycle and Trade. *Journal of International Trade & Commerce*. Vol. 9, No. 2

Yun J.H.J., Mohan A.V. (2012) The Relation Between Internal and External Open Innovation. In Bream A, Tidd J. (eds) *Perspectives on supplier innovation; Theories, Concepts and Empirical Insights on Open Innovation and the Integration of Suppliers*, London: Imperial College Press.

Yun J.H.J., Ryu G.W., (2012) A Study on the Difference of Open Innovation Effect between Modular and Non Modular Firms in Korea. *Asia Pacific Journal of Innovation and Entrepreneurship*. Vol. 6, No. 1

J.J. Yun, *Business Model Design Compass*, Management for Professionals,
DOI 10.1007/978-981-10-4128-0_5

Keywords
Emergence of open innovation • Structure of open innovation • Factors of open innovation of SMEs • Internal open innovation attitude • Product life cycle • Modular • Regional innovation system • Cluster

5.1 Emergence of Open Innovation

Until after the World War II, closed innovation was the dominant paradigm in research and development. But owing to several factors, as listed in Table 5.1, this paradigm could not continue continuously.

Chesbrough described closed innovation as any research project that was made in the firm that moves to the market independently, as illustrated in Fig. 5.1. The possibility of any internal research to arrive at the market is thus very small. In contrast, open innovation expressed that any firm does inside usage of knowledge and outside usage of knowledge, as seen in Fig. 5.2. In open innovation, there are three markets: existing markets, new markets, and markets of other companies. Hence, the possibility of research to arrive at market increases.

Chesbrough's (2006a) conceptual definition of open innovation evolved as shown in Fig. 5.2. There is differentiation between internal technology and external technology base. In addition, three different markets such as other firm's market, our new market, and our current market exist.

We can see that the influx of knowledge is broken down into internal knowledge base and external knowledge base from Fig. 5.2. But research and development (R&D) is also the key factor of innovation in an open innovation condition. In addition, external technical institutes, conferences, paper presentations, and a systematic approach to knowledge and utilization of external sources, such as patent data, are also essential. Chesbrough explains that not only to bring internal R&D to market but also outside-in of excellent technology are essential sustainable innovation. Outward open innovation and inward open innovation can exist together.

Table 5.1 Factors leading to a decrease of closed innovation according to Chesbrough (2003)

Factors	Reasons
Skilled workers	The increasing availability and mobility of skilled workers
Venture capital	An enormous expansion of VC from the 1980s that lured individual personnel from their labs could be lured away by attractive risk/reward compensation packages to join new start-up firms
Using outside knowledge	External options for ideas sitting on the shelf
Outside suppliers	The increasing capability of external suppliers

Source: Chesbrough (2003, pp. 34–41)

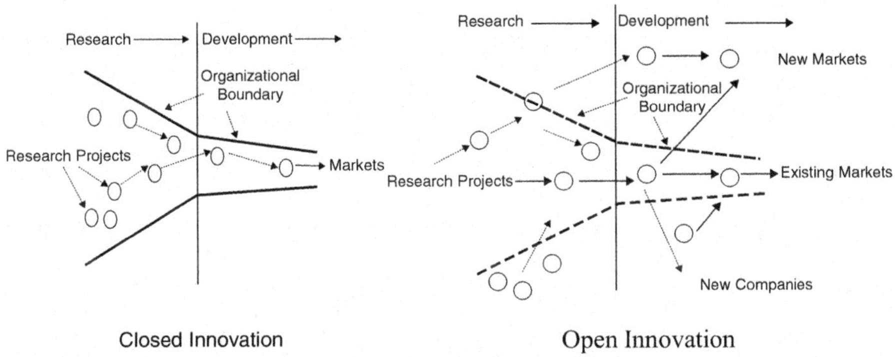

Fig. 5.1 Comparison of closed innovation and open innovation (Source: Chesbrough (2003, pp. 22, 25))

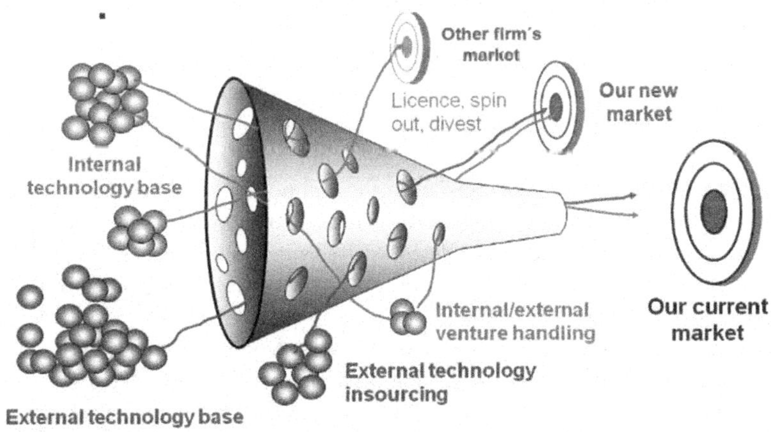

Fig. 5.2 Concept model of open innovation paradigm by Chesbrough (Source: Chesbrough (2007, p. 17))

The core of the open innovation paradigm expresses different roles of three factors, internal research, venture capital, and intellectual property, as outlined in Table 5.2.

First, a new role of internal research is to connect outside knowledge to internal knowledge. Second, a new role of venture capital is to find creative manpower and technology in existing firms and to invite them to new creative firms with innovative motivation. Third, intellectual property is an integral part of an open innovation firm.

The original concept of open innovation by Chesbrough was focused on the firm level (Chesbrough 2003, 2006). But the level of open innovation has been expanded to the sector level, cluster or regional innovation system level, or national innovation system (Cooke et al. 1997). In addition, it was expanded to interorganizational context, knowledge networks and geographic locus, systemic innovation, or value

Table 5.2 The role of three factors in open innovation

Factors	Role
Internal research	Beyond knowledge generation to connection
Venture capital	VC serves as pilot fish for potential market opportunities, because start-up firms are selling real products to real customers, who pay with real money
Intellectual property	Open innovation companies use licensing of IP extensively to create and extend markets for their technology

Source: Chesbrough (2003, pp. 52–57)

networks (Chesbrough et al. 2006, pp. 205, 220, 241, 258). Open innovation platforms, industrial dynamics, and R&D projects are other examples of the diversification of open innovation levels (Chesbrough et al. 2014, pp. 71, 94, 115).

5.2 The Structure of Open Innovation

Open innovation appears in completely different forms, depending on company size and the nature of the industry to which a company belongs. First, in the case of big companies, the degree of accumulation of internal technological capabilities and resources is high (Cui et al. 2012). On the other hand, small- and medium-sized enterprises (SMEs) have remarkably inferior technological capabilities because of shortages of human and material resources and also a lack of accumulated technologies or experience in production. These differences related to enterprises' scale result in considerable differences in the structure of open innovation (Sun and Wang 2011).

Further, the structure of open innovation varies depending on whether the industries to which enterprises belong are manufacturing or service, because the nature and contents of the knowledge required in processes by the stage of innovation are different between these industries. Unlike manufacturing industries, in the case of service industries, not only the stages of research and development and idea excavation but also the stage of product sales is directly related to product development. Thus, the room for open innovation is very large in all these stages (Chesbrough 2011).

Based on a literature review, the structure of open innovation according to size of firm and whether it belongs to an industry related to either manufacturing or service can be categorized into four types, as shown in Fig. 5.3. Big firms in the manufacturing industry have traditional open innovation channels (Chesbrough 2003, p. 24). Diverse technologies in R&D steps from in and outside of the firms can be invited. And diverse markets from modern markets, through expanded markets to new markets, can appear from this type of open innovation. But in the marketing phase, the level of open innovation will decrease more than in the R&D phase.

Large firms in the service industry seem to be very active in open innovation at the R&D phase and marketing phase together, because for firms in the service industry, the marketing phase itself is a kind of product. Firms in the service industry try to develop new products from R&D and from marketing processes.

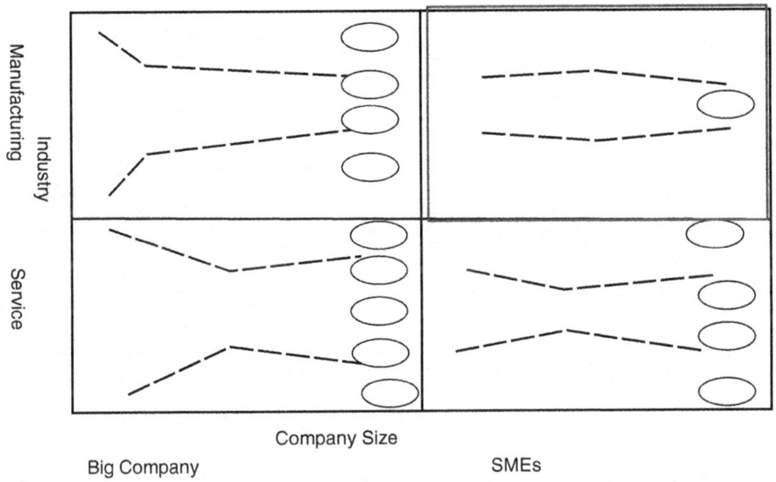

Fig. 5.3 The structure of the open innovation model (Source: Yun and Jung (2013))

SMEs in the service industry carry out similar open innovation to large firms in the service industry. But they do not have enough resources to conduct diverse open innovation. The extent of open innovation by SMEs in the service industry is consequently not as active as that of the larger firms.

Figure 5.3 shows the difference in the structure of open innovation among big companies in manufacturing, big companies in service, SMEs in manufacturing, and SMEs in service. This structure is characterized by the fact that open innovation is activated in different phases of product development and production rather than in the beginning phase for innovation; that is, the phase of research and development and that of open innovation is consistently implemented presuming the creation of certain products for existing markets rather than aiming at the creation of new markets. This structure shows that SMEs' needs for external technologies are related with product development and production. This conflicts with the situation that technology transfer by universities and research institutes is focused on research and development itself or pure new technology.

5.3 Factors of Open Innovation of SMEs

SMEs actively implement open innovation in all stages of product life cycles (Yun 2015). In particular, in the case of SMEs belonging to the manufacturing industry, first, the acquisition of ideas, knowledge, or technologies from external sources has quite important meaning from the stage of defining market requirements, as in Fig. 5.4. Open innovation is important for defining market requirements that will enable them to concentrate their resources to compete and differentiate in situations where large companies or competitor enterprises already exist. In fact, the entrepreneurship role of the CEOs of SMEs plays a very important role in this phase. This

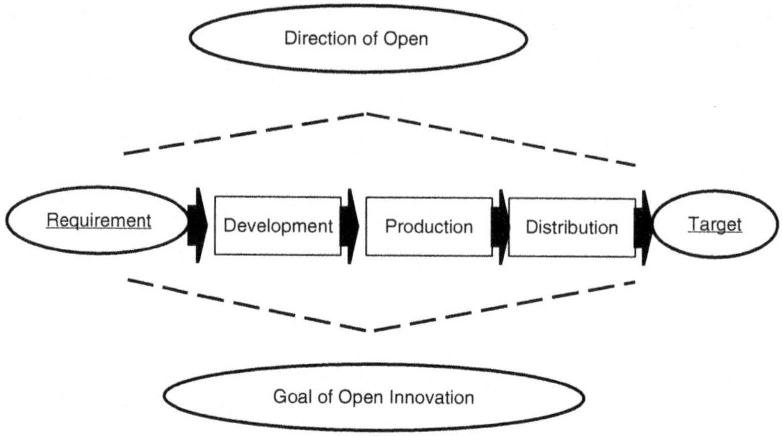

Fig. 5.4 Key factors of open innovation model of SMEs in manufacturing industry (Source: Yun and Jung (2013))

is in contrast with large companies where individuals cannot develop new business models easily (Chesbrough 2007).

Second, SMEs in the manufacturing industry pioneer new niche markets within existing markets and make target markets clear throughout the entire processes of innovation beginning from the initial stage of open innovation. Unlike in the case of large companies, open innovation of SMEs is characterized by the fact that target markets are clearly defined and included in existing markets and that the number of target markets is one or limited to a small number (Sun and Wang 2011). Target markets actually become the ground for determining the range, direction, and contents of open innovation strategies.

Third, open innovation of SMEs in the manufacturing industry is mostly concentrated on inward open innovation. SMEs concentrate their efforts on securing the technology, knowledge, and ideas necessary in all processes of developing products for a specific target market. In emergency situations where SMEs cannot sell products they have developed in markets due to competition with large companies or otherwise, they attempt outward open innovation with the technologies they have developed for the relevant target markets. However, in this case too, the same factor that interrupted the sales of the relevant products in the markets will again become a factor to prevent outward open innovation.

Fourth, rather than aiming at securing creative new technologies, SMEs in the manufacturing industry appear to pursue open innovation in order to reduce costs (Van de Vrande et al. 2009). The goal of open innovation of SMEs in the manufacturing industry is the acquisition of knowledge, technologies, or ideas related to the methods to develop and produce the cheapest products that can satisfy market requirements for defined target markets.

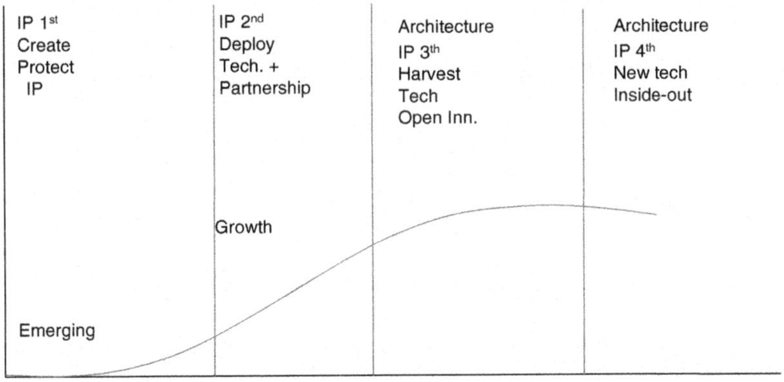

Fig. 5.5 Four product life cycle and the IP life cycle model (Source: Chesbrough (2006a, pp. 90, 92–93))

The following section presents the details of a case study of a Korean SME from the manufacturing sector in order to reflect the above issues of the open innovation structure.

5.4 The Difference of Open Innovation According to Product Life Cycle

5.4.1 Product Life Cycle

As illustrated in Fig. 5.5, Chesbrough divided the product life cycle (PLC) into an emerging stage, growth stage, maturity stage, and decline stage. He also divided the relevant intellectual property life cycle model into the following: "the initial stages of a new technology," where the technology is secured and its protection system is provided; "the next phase," where the technology is applied to manufacturing and internal technological capability is promoted through partnership; "the third phase," where the industry can obtain profit positively through open innovation (OI); and finally "the final phase," where the inside-out open innovation through which obtained profit overlaps with the outside-in open innovation strategy in order to secure new and alternative technology.

From the perspective of IP management strategy, Chesbrough showed that the industry needs to maximize its profits by positively utilizing its technologies through a more direct open innovation when the technology reached its maturity stage.

If we divide product innovation and process innovation, such as in Fig. 5.6, life cycles of the two are different from each other. During the early stage of a life cycle, product innovation can be bigger than process innovation. But in the late stage of a life cycle, process innovation is bigger than product innovation. The sources of innovation are provided from various subjects as well as from the inside in the overall stages of the product life cycle (Utterback and Abernathy 1975).

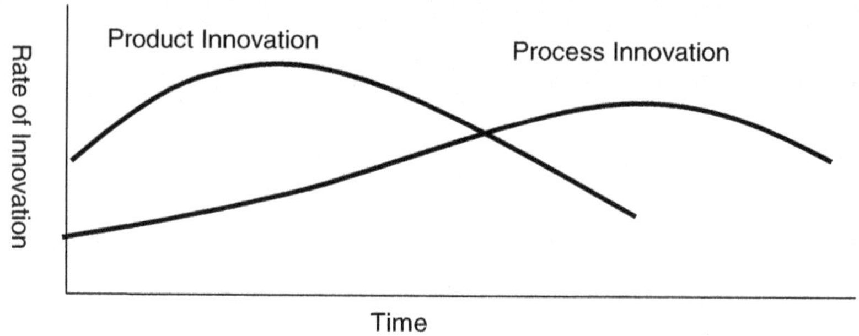

Fig. 5.6 Utterback–Abernathy model in product maturity and innovation (Source: Utterback and Abernathy (1975, p. 639))

Also, according to the product life cycle, the location of the idea source of open innovation is different. That is, they argue that open innovation in the emerging and growing stages results from external users' need; in the mature and declining stages, new and innovative ideas result mainly from the internal developer of the product (Ettlie 2006, p. 256).

5.4.2 Level of Open Innovation According to the Product Life Cycle at Sectorial Innovation System Level

The technological and market uncertainty in any sector at the beginning of the product life cycle is usually accompanied by increased knowledge exchange between firms in which experts develop a common language, which is not yet codified (Rosenkopf and Tushman 1998; Rowley et al. 2000). It is also common practice to interact with a wide variety of external factors, such as customers, other firms, suppliers, research institutes, etc., to strengthen the installed base of the firm's technology (Ozman 2011). It argues that in the beginning of a sector life cycle, open innovation policies usually serve the dual purpose of exploring distant knowledge sources and exploiting potential network effects to strengthen the installed base of a technology. But in the later phases of the sector life cycle, after the emergence of a dominant design, open innovation incentives and effects depend largely on the product system architecture (Ozman 2011). In the mature phases of the industry life cycle, a dominant design emerges, and firms gain familiarity with the dominant design and complementary products, which support the existing dominant design, emerge (Abernathy and Utterback 1978).

Through previous studies by Chesbrough, it can be expected that a considerable relationship exists between product life cycle and open innovation. In the case studies, it was shown that open innovation is more actualized when a sector grows into new areas, and investment increases (Chesbrough 2003); examples include when IBM advanced into the system software area and reinforced its new capability

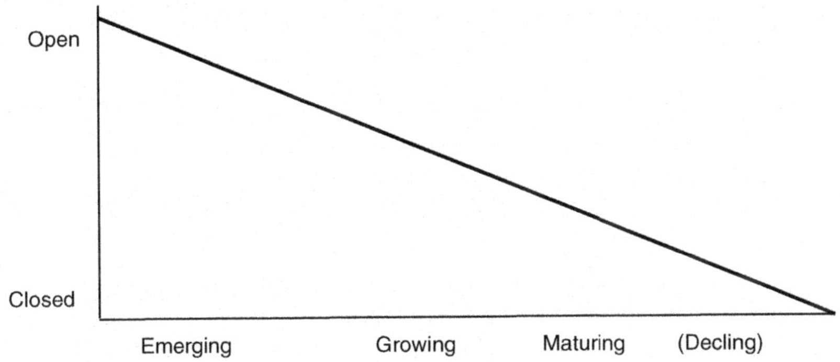

Fig. 5.7 Relationship between product life cycle and level of open innovation in the sectorial level (Source: Yun et al. (2013))

beyond the existing computer area and also when Intel continuously developed new areas, such as the flash memory area.

Microsoft's operating system was a derivative of one created by another company that Microsoft obtained a license from, and many of the major software applications Microsoft created were based on products invented elsewhere, such as Google, which is spending a lot of its money to buy new mobile platforms from all over the world, even though it is now a global top firm in that sector (Brandt 2009, p. 194). Not only Microsoft but Google also did not hesitate to adopt "outside-in" open innovation in order to create new software or new products. In fact, a considerable part of the success of these sectors depended on open innovation. In particular, it cannot be underestimated that Google is trying to create a profit model through maximized utilization of external knowledge and technology of all individuals and sectors through the maximization of open innovation of the entire world.

Based on several cases and literature reviews of product life cycle, we can establish a research model for the relationship between product life cycle and level of open innovation at sectorial innovation system, as shown in Fig. 5.7. Figure 5.7 shows that the level of open innovation has an inverse relationship with product life cycle in the sectorial level. That is, open innovation mainly occurs in the emerging and growing stage product life cycles at the sectorial innovation system. Furthermore, when the product life cycle reaches the maturity and declining stages, closed innovation will lead the situation, surrounding the relevant product's technology at the sectorial innovation system.

The mobile industry is considered to be in the growth stages compared to the automotive sector. However, it is closer to the maturity-stage sector compared to the new-growth sector such as the IT medical care subsector, IT robot subsector, solar energy subsector, and fuel cell subsector.

The open innovation level of emerging and growing stage sectors and that of post growth-stage sectors are different in open innovation with university channels. The level of open innovation with the university channel will decrease when the product life cycle grows from the emerging stage (new-growth sector) through the growth sector (mobile sector) to the mature stage (mobile sector).

5.4.3 Difference of Open Innovation Effect According to the Product Life Cycle at the Firm Level

The breadth and depth of open innovation do not unilaterally increase the firm's performance, but they have an inverted U-shape relationship (Laursen and Salter 2006). There are several managerial factors promoting firms' performance through open innovation, such as those included in the establishment of extensive networks of interorganizational relationships with a number of external actors, in particular, universities and research institutions, organizational systems focused on accessing and integrating the acquired knowledge into the firm's innovation processes, new evaluation criteria to focus more on external sources of innovation, and the use of knowledge management systems, which are able to support the diffusion, sharing, and transfer of knowledge within the firm and within the external environment (Chiaroni et al. 2010).

In the meantime, as shown by Chesbrough in his IP management strategy on the product life cycle or by Utterback in his sources of innovation based on the product life cycle, the open innovation effect is expected to exist in all stages of the technology life cycle (Chesbrough 2003; Ettlie 2006; Utterback and Abernathy 1975). Even in the developing process of a business model, there exists an open innovation effect in all steps of the open business model (Chesbrough 2006a, p. 111). As shown in Fig. 5.8, there is an open innovation effect in all stages of the product life cycle from the emerging and growing stages to the maturity and declining stages at the firm level. This means that open innovation of any firm produces innovative performance. Commitment to learning, shared vision, and open-mindedness is positively related to product innovation performance (Yoon 2004, p. 86). This means that open innovation has a high positive effect on every stage of the product life cycle to the performance of the firm.

Open innovation shows a statistically significant effect not only on growth stage firms but also on mature stage firms.

5.4.4 Rethinking the Relationship Between Open Innovation and Product Life Cycle

Classical PLC theory argues that many new products are initially discovered and produced in developed countries and exported to developing countries, but as technology for production becomes more standardized, production shifts to developing countries due to lower labor costs (Vernon 1966). In the PLC theory, the rate at which an individual firm discovers and successfully markets new products is either treated as exogenously given (Krugman 1979) or as a "deterministic function of the firm's expenditures on new product development" (Jensen and Thursby 1986). In

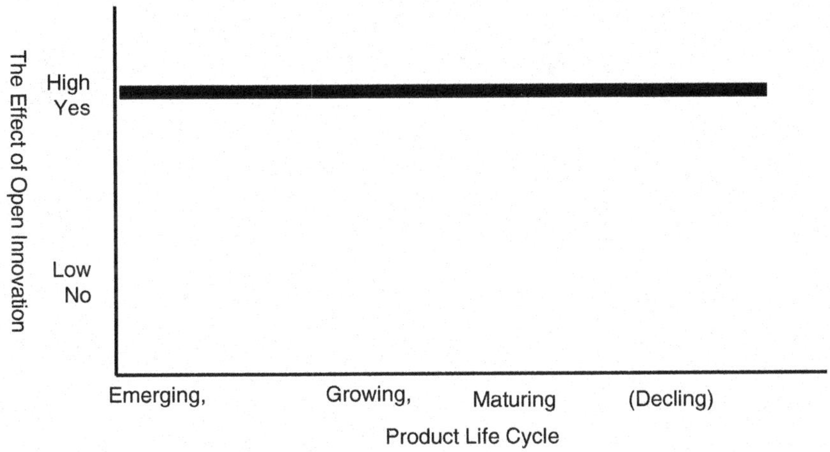

Fig. 5.8 The relationship between the effect of open innovation and PLC in the firm level (Source: Yun et al. (2013))

contrast, Schumpeter stressed that firms compete with each other to successfully introduce new products (Schumpeter 1942).

Sustained product innovation in developed countries enables workers to earn higher wages than comparable workers in the in developing countries (Segerstrom et al. 1990). According to the PLC theory, patent protection or internal R&D triggers an increase in R&D cost and a decrease in the revenue of developed countries. Hence, if open innovation is activated in developed countries for product innovation, revenues of products will be extended. Congruently, under the knowledge-based economy, most products will fall into a commodity trap, after a product is introduced to the market (Chesbrough 2011, p. 124). Thus, an ordinary introducer of new products in a developed country cannot continue to obtain revenue from the product. Three kinds of commodity trap, the deterioration trap, the proliferation trap, and the escalation trap, decrease the PLC of new and innovative products and diminish the competitive advantage of the firm dramatically (D'Aveni 2010, p. 56; Lina and Dalim 2013). In today's rapidly changing business landscape, new sources of sustainable competitive advantage can often only be attained from business model reinvention that is based on disruptive innovation and not on incremental change or continuous improvement (Voelpel et al. 2004).

Sustained product innovation by open innovation will maintain the revenue of the product for developed countries. The Vernon effect from the PLC is not manifested in several developed countries presently, but the open innovation effect for escaping from commodity trap is sure.

The structural relationship among open innovation, commodity trap, PLC, and trade is summarized in Fig. 5.9 based on open innovation theory, the commodity trap concept, and product life cycle theories. If open innovation is actualized, the product of the firm does not fall into the commodity trap. A postponed or prevented fall into a commodity trap motivates the decrease of the product life cycle. The

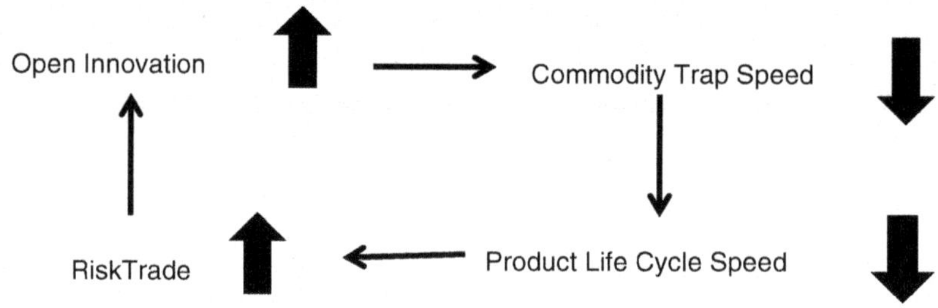

Fig. 5.9 The relationship between open innovation, commodity trap, PLC, and trade (Source: Yun et al. (2013))

decreasing speed of the product life cycle triggers an increase in trade, and increased trade actualizes open innovation through the trade process itself.

Nowadays, the requirements for reinvention of capitalism are growing. Creating shared value (CSV) is integral to a company's profitability and competitive position (Porter and Kramer 2011). The main way to acquire CSV or dynamic energy of sustainable development in or out of firms is open innovation (Christensen 2012).

5.5 The Role of Internal Open Innovation Attitude

5.5.1 Internal Open Innovation Attitude

While open innovation has two sources—internal or external technology—there is an emphasis on the internal attitude or culture of firms for open innovation. There are eight points of differentiation for open innovation. Two of them are as follows: (i) the centrality of the business model in converting R&D into commercial value and (ii) the proactive and nuanced role of intellectual property (IP) management, which is directly related with the internal open innovation attitude of a firm (Chesbrough 2006b, p. 11). The argument is that any firm that has an open attitude or culture to new ideas and new innovation can easily actualize new business models and intellectual property. In addition, there is a new role for internal R&D as a knowledge connector and broker, and this "open innovation thinking" means having an internal open innovation attitude or internal open innovation. Open innovation thinking changes the role of the research function. It expands the role of internal researchers to include not just knowledge generation but also knowledge brokering (Chesbrough 2003, p. 56). Thus, in the conceptualization of open innovation, there is external open innovation and internal open innovation, and these two aspects affect each other.

5.5.2 Deep Relation Between Internal Open Innovation Attitude and External Open Innovation

Apple Corporation built a creative culture under the leadership of Steve Jobs. He organized management retreats as an annual ceremony. Anyone who attends a retreat can take three or four full days to develop creative ideas. In addition, he gave any team that was involved in new innovative projects such as the Macintosh, iPhone, and iPad big compensation (Young and Simon 2005, p. 58). This policy gave his company an "open attitude" to new ideas. And this internal open attitude has a relation with its external open innovation.

At Apple in the early days, Steve Jobs would go outside the company for searching creative product design and advertising. Steve surely denied the "NIH" philosophy—"not invented here." The technology had to be created within Apple; if his technical wizards did not know how to do something, they would just hire someone who did. Going outside the company simply was not acceptable. Someone who can't change their ideas is a prisoner of his past. Steve broke out of that prison (Young and Simon 2005, p. 279). Several of Apple's products such as the Macintosh, iMac, and iPad were developed through internal creative and open research, but the initial ideas were from outside Apple. Some of the technologies took root through the arrival of key employees at Apple. The Macintosh computer embodied many of the user-interface design concepts created at Palo Alto Research Center (Chesbrough 2003, p. 5).

Google has a 20% rule, which means that any employee can participate in any interesting research or technology job on his or her own volition (Luoyaozong 2005, p. 111). This rule allowed employees to develop several creative new products such as the Google Deskbar, Google Books, Google News, and Google Alerts. It also led Google to have an "internal open attitude" to new products. In addition, Google recruits creative people from all over the world. As a result, it can be argued that the "internal open innovation attitude" has a deep relation with external open innovation. 3M also has its 15% rule, which is similar to Google's. This rule was behind the creation of the Post-it. It also spawned 3M's adhesive tape business, which currently produces more than 700 specialized products for medical, electrical, home, and industrial applications (Luecke 2003, p. 5).

In addition to the cases discussed thus far, managing open innovation in a world of intermediate markets for ideas also requires the construction and support of a rich internal innovation network (Chesbrough 2006a, p. 20). Indeed, an internal innovation network is essential for the open innovation model and markets such as Intellectual Ventures, InnoCentive, and Ocean Tomo. In conclusion, internal open innovation attitude and external open innovation are deeply related. This leads to the conclusion that there is a deep relation between internal open innovation and external open innovation.

5.5.3 The Function of Internal Open Innovation Attitude to External Open Innovation

The inventors of a new technology are often not the first to profit from that technology, and one of the reasons for this is the difficulty of transferring new research discoveries into production (Chesbrough 2003, p. 115). Intel put together an R&D group and a production group in such a way that every new researcher has to work on production from the beginning of their career. In addition, Intel invested in standardizing its equipment between "lab" (laboratory) and "fab" (fabrication). These policies allowed the company to develop an internal open attitude. Intel's research philosophy fostered an external orientation to the generation of knowledge (Chesbrough 2003, p. 130). Indeed, Intel's internal open innovation attitude triggered several mechanisms to access external innovation such as the *Intel Technical Journal*, the funding of about three hundred external research projects and lablets.[1]

Knowledge creation capability rather than organizational knowledge is a critical resource in an organization (Nonaka and Konno 1995, p. 71). Not merely core competences but also capability, which is used as the know-how at the organization level, is considered better for open innovation. Nonaka's knowledge creation theory asserts that the interaction between tacit knowledge and coded knowledge creates knowledge in both quantity and quality (Nonaka and Konno 1995, p. 93). The knowledge creation capability means internal open innovation, and it is one of the main resources for knowledge development in the interaction between tacit and coded knowledge. Knowledge firms, which are similar to active external open innovation firms, have several organizational cultures such as an open culture and the not-quantity-but-quality culture, and they stress the importance of communication (Nonaka and Konno 1995, p. 204). According to Nonaka, an internal open attitude also increases external open innovation. That is, an internal open innovation attitude will have a modulating effect on the relation between external open innovation and innovative performance. Here "modulate" means "extend," "increase," and "expand". If we analyze the relation between internal attitude and open innovation, we can understand more comprehensively the open innovation process and create open innovation strategies.

[1] "Lablets" are small research facilities located adjacent to three leading university research centers—Carnegie Mellon University, the University of California, Berkeley, and the University of Washington —instead of next to Intel fab facilities. As with other parts of its R&D system, Intel manages these new entities in a decidedly untraditional manner (Chesbrough 2003, p. 123).

5.6 The Difference in Open Innovation Depending on Whether a Firm Is Modular or Not

5.6.1 The Relation Between Module and Open Innovation

Many of the technology-based products are inclined to become modularized as the products become more complex and sophisticated. While the extent of modularization differs depending on the technical maturity or the industrial maturity, most of the technology-based products are inclined to be modularized in order to minimize the influence from outside and interactional uncertainty resulting from technical innovation of specific parts. Compared to interdependent architectures with high uncertainty resulting from interactions between them, modular architectures imply a system that minimizes the interdependence between parts by minimizing the interdependence between architectures. Modularization means that the proportion of modular architecture increases among various architectures. The greater the tendency to modularization an industry has, the easier it is for the industry to adopt new technical innovation results from the outside and optimize them. In short, as the proportion of modular architecture of the technology-based products increases, outside-in open innovation promotes the performance of the industry. In any early age of a technology's evolution, there are many possible ways that the different component technologies might relate with one another: the greater the number of components, the greater the number of possible interconnections between them (Chesbrough 2003, p. 58). When many interactions inside the product's system are not known, it is possible for the industry to make new architecture needed for the product through R&D. Under the uncertain and complicated situation, however, it is likely that the industry will fail when relying on external technology to solve the interconnection problem. That is because we rely on external technology with the characteristics and details of the product completely unknown, and it may be possible to produce different partial products with different functions depending on the industry that applies the technology to the production. Thus, the industries should completely understand how the technology works, accumulating experience for a long time so as to adjust the complexity and to solve the ambiguity of the technology (Chesbrough 2003, p. 59).

5.6.2 The Difference in Open Innovation in Interdependent or Modular Architecture

Chesbrough (2003, p. 60) distinguishes Interdependent Architecture in Fig. 5.10 from Modular Architecture in Fig. 5.11 according to the interrelation between parts comprising technology-based products and between parts and system. He argues that when new technology is introduced, technology-based parts are primarily made from inside due to the great uncertainty and complexity, and they assume the forms of interdependent architecture, based on the high interaction between parts. Further, in the case of this interdependent architecture, the internal R&D plays an important

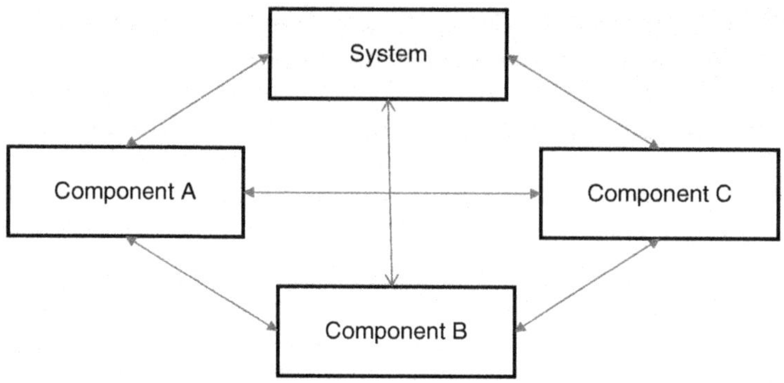

Fig. 5.10 Interdependent architecture model (Source: Chesbrough (2003, p. 60))

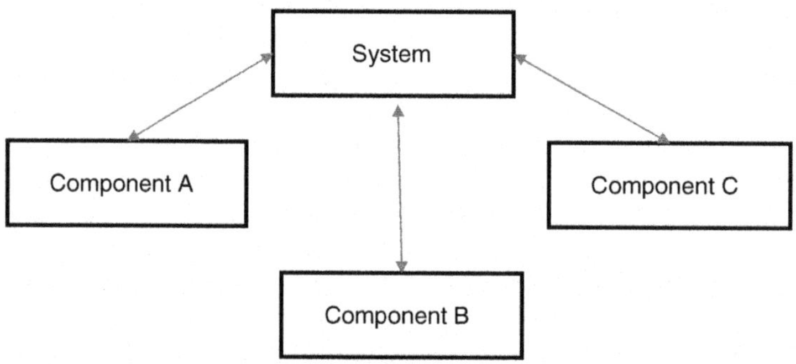

Fig. 5.11 Modular architecture model (Source: Chesbrough (2003, p. 61))

role in understanding the relation between parts and in adding new architectures. Under the interdependent architecture like this, specific parts industries want their parts to occupy an important position and to control and manage the whole system, and they try to do so.

But more valuable architecture makes it possible for other subjects (industries) to utilize their expertise and to have an opportunity to participate in the system, diminishing the interdependence and decreasing the complexity between architectures. A need for effective connection implies that the industries cooperate and compete with other industries in the system. As the technology advances, however, and as the function and characteristics of the technology become clear, the availability of the technical management also increases, and then the industries try to limit and clarify the function and boundary of the product.

In addition, as for each vendor providing parts, they can more easily add themselves as a vendor to provide specific parts according to whether it meets their requirements and expectations or they may drop existing vendors. Furthermore, with the advancement of modular architectures, intermediate markets also develop, industries that produce parts relating to specific parts of the architecture emerge, and external technology competes with partitions of existing architecture (Chesbrough 2003, p. 61).

In this well-established modular architecture, it becomes possible for hundreds or thousands of firms to pursue various technical innovations without worrying about the possible impacts of their improvements on the other parts of the system. Industries pursuing open innovation find it easier to actively innovate in modular architecture and to reflect it as a new part of the system, and they voluntarily take part in intermediate markets in modular architecture (Chesbrough 2003, p. 62). Therefore, in modular architecture, it becomes possible for the industries to save time and expense and to make more innovative products by purchasing specific parts from outside rather than making them independently.

While the prior research on products and technology regarded specific product technology as isolated and independent, most of the current research shares the technology including intellectual property by connecting suppliers, consumers or competitors on the value chain (Chesbrough 2006, p. 87). The technology connection shown in Fig. 5.12 can lead to another form of patent infringement problem regarding intellectual property due to customers' or suppliers' standpoints other than industries' themselves.

As described in Fig. 5.12, most technology-based products, particularly modular architecture products, are supplied with technology including IP following the value chain, and in the case of industries producing intermediate goods, open innovation

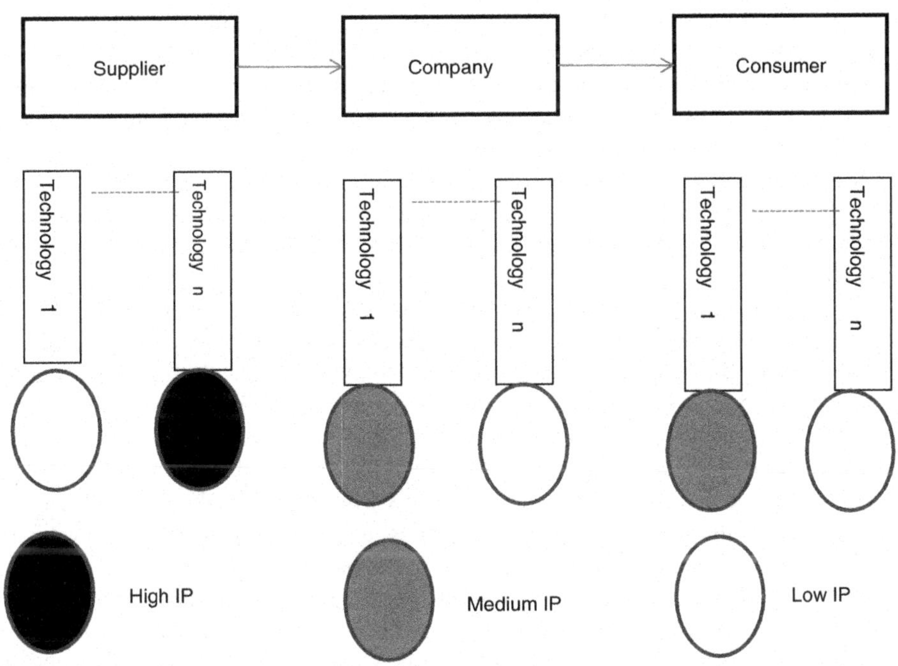

Fig. 5.12 A patent map of the value chain (Source: Chesbrough (2006, p. 87))

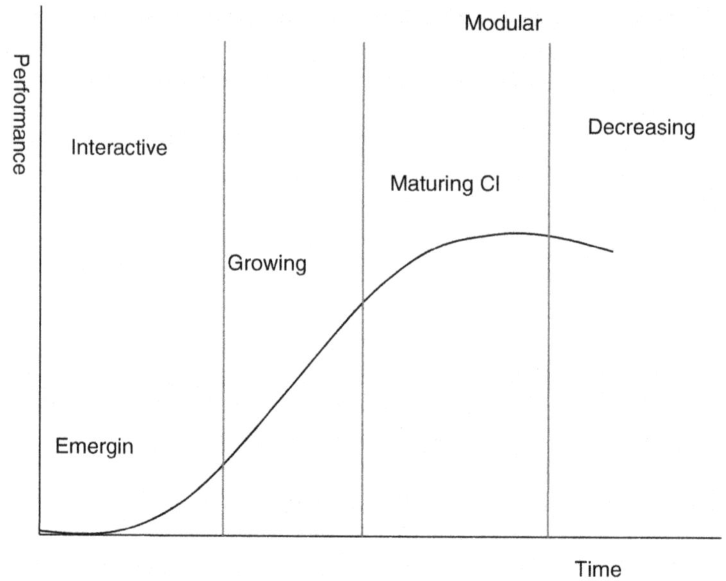

Fig. 5.13 The match between TLC, and modular or interactive architecture (Source: Chesrough (2006, p. 90) modified)

inevitably occurs, which in turn provides their technology to other customer industries. With regard to modular architecture products, as it is possible for them to optimize technical innovation through open innovation on the value chain while minimizing the influence on other parts in the system, they pursue outside-in open innovation from industries whose technology and ideas are superior.

Meanwhile, individual technology or technological systems have a life cycle as shown in Fig. 5.13. That is, starting from the merging stage when the technology appears for the first time, it goes through a growth stage, maturity stage, and declining stage.

The emerging stage, just before the dominant design appears, is when the various technical feasibilities are being tested. Taking the automobile for example, this is when the gasoline engine of the automobile fought fiercely with the steam engine and electric engine to obtain the status of dominant design (Chesbrough 2006, p. 91). The growth stage is when the technology is rapidly progressing after the dominant design has been set and this is when Ford's model T and A were explosively produced and sold. The maturing stage is when the technology is well known and the use of that technology reaches its peak; an example is when sports vehicle and mini-vans were produced and supplied through segmentation of the automobile market. The declining stage is when alternative products to existing products are being made and established in the market with the turnover of existing products declining; the decline of existing car tires in the USA due to the advent of European car tires is a good example (Chesbrough 2006, pp. 91–92).

In modular architecture, it is possible for companies to provide parts based on optimal technology of the module. Namely, in modular architecture, it is possible for companies to achieve an optimal innovation effect through outside-in open innovation. Accordingly, the open innovation effect appears to be far greater in the modular-based industry than in independent architecture-based industries. As the technology matures, interdependency between parts decreases, and independent architecture is transformed to modular architecture. In the meantime, while the combustion engine-based automobile industry is currently in the mature stage after passing through the emerging and growth stages, the smartphone industry is maintaining a rapid upswing.

5.7 The Difference in Open Innovation According to Regional Innovation Systems or Clusters

Regional innovation is enabled by the exchange of knowledge among a diverse set of institutions and organizations. Accordingly, an optimal open innovation strategy would exploit multiple types of links to multiple types of institutions, and each type of association would result in the flow of different pieces of knowledge (Simard and West 2006, p. 226). Thus, a cluster that belongs to a regional innovation system has several key institutions such as high-quality research universities, active venture capitalists that serve as a "powerful institutional force," flagship firms that act as a breeding ground for knowledge creation and further ventures, and industry-specific innovation organizations such as public research institutes for biotechnology.

Therefore, market failure resulting from the exchange of asymmetric knowledge between clusters or regional innovation systems requires open innovation (Cooke 2005, p. 81). Considering the main bio-clusters in the USA, such as Boston, San Francisco, and San Diego, we can easily identify bioregional knowledge asymmetries by using variables such as scale, sector, and space, as shown in Table 5.3.

This kind of open innovation has spread from biotechnology and post-genomics to electronics, energy, and even homecare products, as seen in the case of Procter & Gamble's replacement of its R&D division by a C&D (connect and develop) division (Cooke 2005, p. 84). However, in order to maximize the innovation capability of any cluster and strengthen the global competitiveness of firms, the innovation processes at the regional level that accelerate global open innovation will need to be strengthened.

The relationship between open innovation and a cluster is illustrated in Table 5.4. If a cluster comprises a diverse set of industries and adopts an open innovation strategy, the firms belonging to that cluster will have a greater potential to develop.

One of the main reasons why firms gather around a cluster is that they can easily access knowledge and information, both of which are essential for their survival. Some recent cluster studies have emphasized how both local activity and the economic growth rate increase when the firms in a cluster benefit from information that is easily available from potential partners in the vicinity and, perhaps more importantly, as a result of the ease with which they can conduct business with such local

Table 5.3 Bioregional knowledge asymmetries, domains, capabilities, and innovation systems

	Scale	Sector	Space
Exploration	"Big pharma"	Biotechnology	Boston
Examination	Screening	Genomics	San Francisco
Exploitation	Drug	HIV/AIDS	San Diego

Source: Cooke (2007, p. 26)

Table 5.4 Characterization of successful and potentially successful knowledge clusters[a]

	Specialization	Diversification
Pipeline	1. Embryonic	4. High success
Open science	2. Innovative	3. High potential

Source: Cooke (2005, p. 93)

[a]Pipeline industries in which knowledge cannot be transferred to outside the company are akin to a closed innovation system; on the other hand, a situation in which knowledge can be transferred either inward or outward is similar to an open innovation system

firms (Maskell 2001). Firms that belong to any cluster can acquire useful information and knowledge from similar firms along the horizontal axis and from suppliers and consumers along the vertical axis.

Many of the existing theories related to the innovative activity of clusters focus on the external effect, which is anything that raises the return to a particular firm located in a region as a result of the location of other firms in the same region (Bresnahan et al. 2005, p. 115). In other words, an external effect implies that firms located in a cluster can learn about markets and technical developments from their colleagues in neighboring firms. However, cluster formation is a process that relies on the co-evolution of technology, business models, and local supporting institutions and, therefore, appears to be more prevalent in prepared regions (Feldman and Braunerhjelm 2006, p. 11).

When a major technological innovation occurs, new clusters appear and become the locus of the new activity; over time, new industries develop and clusters are formed. As a result of the interactions between agglomeration economies and diseconomies on the one hand and incremental versus radical innovation on the other, these new industries may eventually lose their advantage (Maggioni 2006, p. 219). In conclusion, if the external effect of a cluster is positive, the net number of new firms will increase. Conversely, if the external effect of a cluster is negative, the net number of new firms will decrease. Further, if open innovation is implemented along with an increase in the net number of new firms, the cluster will develop in a manner as illustrated in Fig. 5.14. But, if new firms do not join the cluster, the cluster will eventually shrink in quantity and quality.

According to Saxenian and Hsu (2005, pp. 235–260), the Silicon Valley–Hsinchu Connection served as the main trigger for the growth of the Hsinchu cluster. A small group of Taiwanese immigrants set up a local branch of the Chinese Institute of Engineers (CIE), which is commonly regarded as the "grandfather" of the Chinese

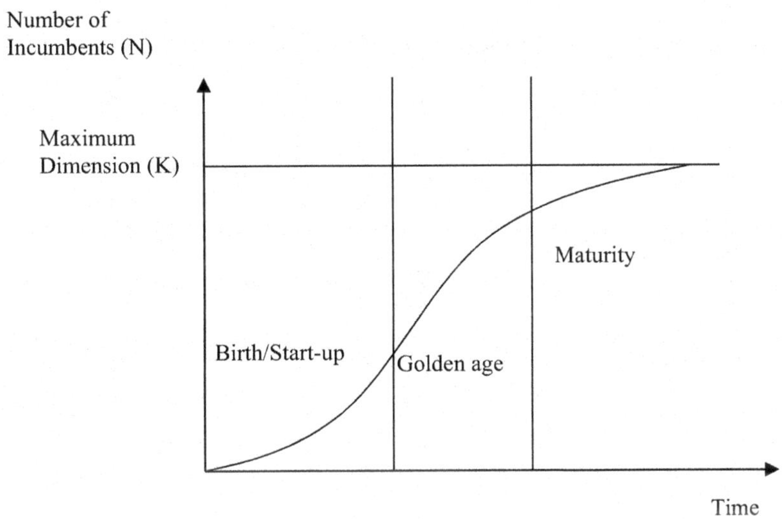

Fig. 5.14 Development of an industrial cluster, in isolation (Source: Maggioni (2006, p. 223))

professional organizations in the Silicon Valley, and Taiwan's policy-makers unwittingly supported the extension of Silicon Valley's Chinese network to allow them to work together with their counterparts in Taiwan. In the end, frequent advisory meetings and technical interactions supported the creation of personal and professional relationships between engineers, entrepreneurs, executives, and bureaucrats on both sides of the Pacific (Saxenian and Hsu 2005, pp. 243, 245). Many firms in the Hsinchu cluster that had a global open innovation network, particularly with Silicon Valley, developed rapidly from small industrial companies to global companies in high-tech industries such as semiconductors or display units.

Research Question

1. Select any firm and analyze its structure of open innovation by the knowledge funnel introduced in this chapter.
2. Select any firm and analyze the differences in open innovation according to whether the firms are modular or not, product life cycle, and location.
3. Select more than two firms and compare their internal open innovation attitude and its effects.

References

Abernathy WJ, Utterback JM (1978) Patterns of industrial innovation. Technol Rev 80(7):40–47
Brandt RL (2009) Inside Larry and Sergey's brain. Penguin Group, New York

Bresnahan T, Gambardella A, Saxenian A (2005) Old economy inputs for new economy outcomes: cluster formation in the new Silicon Valley. In: Breschi S, Malerba F (eds) Clusters, networks and innovation. Oxford University Press, Oxford
Chesbrough H (2003) Open innovation: the new imperative for creating and profiting from technology. Harvard Business School Publishing Corporation, Boston
Chesbrough H (2006a) Open business models: how to thrive in the new innovation landscape. Harvard Business School Press, Boston
Chesbrough H (2006b) Open innovation: a new paradigm for understanding industrial innovation. In: Chesbrough H, Vanhaverbeke W, West J (eds) Open innovation: researching a new paradigm. Oxford Press, Oxford
Chesgrough H (2007) Business model innovation: It's not just about technology anymore. Strateg Leadersh 35(6):12–17
Chesbrough H (2011) Open services innovation: rethinking your business to grow and compete in a new era. Jossey-Bass, San Francisco
Chesbrough H, Vanhaverbeke W, West J (2006) Open innovation: researching a new paradigm. Oxford University Press, Oxford
Chesbrough H, Vanhaverbeke W, West J (2014) New frontiers in open innovation. Oxford University Press, Oxford
Chiaroni D, Chiesa V, Frattini F (2010) Unravelling the process from closed to open innovation: evidence from mature, asset-intensive industries. R&D Managment 40(3):222–245
Christensen K (2012) Questions for Henry Chesbrough: the man who wrote the book on open innovation explains how not to get stuck in the 'commodity trap'. Rotman Magazine
Cooke P (2005) Regional knowledge capabilities and open innovation: regional innovation systems and clusters in asymmetric knowledge economy. In: Breschi S, Malerba F (eds) Clusters, networks and innovation. Oxford University Press, Oxford
Cooke P (2007) Theorizing regional knowledge capabilities: economic geography under 'open innovation'. In: Suriñach J, Moreno R, Vayá E (eds) Knowledge externalities, innovation clusters and regional development. Edward Elgar, Northampton
Cooke P, Uranga MG, Etxebarria G (1997) Regional innovation systems: institutional and organisational dimensions. Res Policy 26(4):475–491
Cui Z, Loch C, Grossmann B, He R (2012) How provider selection and management contribute to successful innovation outsourcing: an empirical study at Siemens. Prod Oper Manag 21(1):29–48
D'Aveni R (2010) Beating the commodity trap: how to maximize your competitive position and increase your pricing power. Harvard Husiness School Press, Boston
Ettlie JE (2006) Managing innovation: new technology, new products, and new services in a global economy, 2nd edn. Elsevier, Boston
Feldman M, Braunerhjelm P (2006) The genesis of industrial cluster. In: Braunerhjelm P, Feldman M (eds) Cluster genesis: technology-based industrial development. Oxford University Press, Oxford
Jensen R, Thursby M (1986) A strategic approach to the product life cycle. J Int Econ 21:269–284
Krugman P (1979) A model of innovation, technology transfer and the world distribution of income. J Polit Econ 87:253–266
Laursen K, Salter A (2006) Open for innovation: the role of openness in explaining innovation performance among U.K. manufacturing firms. Strateg Manag J 27:131–150
Lina L, Dalin W (2013) The analysis of substituting ability of new products: based on a prey-predator model. J Int Trade Commer 9(1):591–607
Luecke R (2003) Managing creativity and innovation. Harvard Business Essentials, Boston
Luoyaozong Z (2005) The seven success lessons from Google. Xipsaiae Publishing, Seoul. , translated in 2007

Maggioni M (2006) Mors Tua, Vita Mea? The rise and fall of innovative industrial clusters. In: Braunerhjelm P, Feldman M (eds) Cluster genesis: technology-based industrial development. Oxford University Press, Oxford

Maskell P (2001) Towards a knowledge-based theory of the geographical cluster. Ind Corp Chang 10(4)

Nonaka I, Konno N (1995) Intellectualizing capability. Book 21 Publishing Group, Seoul. Translated from Korean in 2009

Ozman M (2011) Modularity, industry life cycle and open innovation. J Technol Manage Innov 6(1):26–37

Porter M, Kramer M (2011) Creating shared value: how to reinvent capitalism and unleash a wave of innovation and growth. Harv Bus Rev 89:62–77

Rosenkopf L, Tushman M (1998) The coevolution of community networks and technology: lessons from the flight simulation industry. Ind Corp Chang 7:311–346

Rowley T, Behrens D, Krackhardt D (2000) Redundant governance structures: an analysis of structural and relational embeddedness in the steel and semiconductor industries. Strateg Manag J 21:369–386

Saxenian A, Hsu J (2005) The Silicon Valley-Hsinchu connection: technical communities and industrial upgrading. In: Breschi S, Malerba F (eds) Clusters, networks and innovation. Oxford University Press, Oxford

Schumpeter J (1942) Capitalism, socialism and democracy. Harper, New York

Segerstrom PS, Anant TC, Dinopoulos E (1990) A Schumpeterian model of the product life cycle. Am Econ Rev 80:1077–1091

Simard C, West J (2006) Knowledge networks and the geographic locus of innovation. In: Chesbrough H, Vanhaverbeke W, West J (eds) Open innovation: researching a new paradigm. Oxford University Press, Oxford

Sun X, Wang Q (2011) Open innovation in small and medium enterprise under the view of knowledge management. Proceedings of the 2011 IEEE ICMIT, pp 4690–4693

Utterback J, Abernathy W (1975) A dynamic model of process and product innovation. Omega 3(6):639–656

Van de Vrande V, De Jong JP, Vanhaverbeke W, De Rochemont M (2009) Open innovation in SMEs: trends, motives and management challenges. Technovation 29(6):423–437

Vernon R (1966) International investment and international trade in the product cycle. Q J Econ 80:16

Voelpel S, Leibold M, Tekie E (2004) The wheel of business model reinvention: how to reshape your business model to leapfrog competitors. J Chang Manag 4(3):259–276

Yoon D (2004) The impacts of Korean exporting firms' market orientation and learning orientation on product innovation performance. J Korea Trade 8(2):85–111

Young J, Simon W (2005) iCon: Steve jobs: the greatest second act in the history of business. Wiley, Hoboken

Yun JJ (2015) How do we conquer the growth limits of capitalism? Schumpeterian dynamics of open innovation. J Open Innov Technol Market Complex 1(1):1–20

Yun JHJ, Jung WY (2013) Open innovation of SMEs in manufacturing from OI structure model. Asia Pac J Innov Entrep 7(1)

Yun JHJ, Ryu GW, Jung WY (2013) Is there any difference in the effect of open innovation according to the product life cycle?: revisiting the relationship between product life cycle and trade. J Int Trade Commer 9(2):17–46

Real Contents and Channels of Open Innovation

6

Abstract

This chapter explores the contents and channels of open innovation (OI) in small and medium enterprises (SMEs) that operate in emerging or growing technological industries in South Korea. Through case studies, this chapter presents concrete contents and channels of open innovation of SMEs. Many studies already have shown the channels and contents of big businesses or multinational companies (MNCs) (Chesbrough, Open innovation: the new imperative for creating and profiting from technology. Harvard University Press, Cambridge, MA. pp. 1–19, 43, 61, 93–112, 113–133; Open innovation: a new paradigm for understanding industrial innovation. In: Chesbrough H, Vanhaverbeke W, West J (eds) Open innovation: researching a new paradigm. Oxford University Press, Oxford, pp. 3–11, 2006a; Open business models: how to thrive in the new innovation landscape. Harvard Business School Press, Boston. pp. 15, 20, 196–203, 2006b). In this chapter, we see concrete contents of open innovation and channels for it in SMEs. Readers will thereupon have an opportunity to conceive their own open innovation contents and channels in their own SMEs. In addition, we see the reality of closed innovation of SMEs. Even though we look at Korean cases of SMEs' open innovation, the target industries are high-tech industries such as the fuel cell industry, intelligent robots, solar energy, and medical instrument industry, all of which will be future industries for both developed and developing countries.

This chapter is mainly based on the following paper.

Yun J.H.J., Mohan A.V. (2012) Exploring open innovation approaches adopted by small and medium firms in emerging/growing industries: case studies from Daegu-Gyeongbuk region of South Korea, *International Journal of Technology Policy and Management*, Vol. 12, No. 1

J.J. Yun, *Business Model Design Compass*, Management for Professionals,
DOI 10.1007/978-981-10-4128-0_6

Keywords
SMEs • Channels of open innovation • Contents of open innovation • Fuel cell industry • Medical instrument industry

6.1 Introduction

How is open innovation being practiced by small and medium enterprises in emerging or growth stage industries? What kinds of channels are used by small and medium firms in the emerging or growth stage? (Yun 2015; Yun et al. 2015)

Using a case study methodology with the data gathered from interviews, we examine real open innovation channels and contents of medium and small firms and their performance in technological intensive industries such as IT medical care instruments, intelligent robots, solar energy, and fuel cells. A user innovation study in medical equipment had been done by case studies and interviews (Von Hippel 2005). Different from statistical analysis of SMEs, this chapter focused on cases of SMEs' open innovation (Laursen and Salter 2006; Yun 2008, 2009). A "semi-standardized interview" was adopted, which proceeds in accordance with prearranged content and the order of a questionnaire, but with the possibility that the interview can be conducted very flexibly regardless of the content and order depending on the situation (Kim and Mauborgne 2004).

In the case of open innovation industries, this chapter examines the practice of outside-in open innovation and inside-out open innovation in small and medium firms in South Korea. In the case of closed open innovation industries, this study analyzes internal R&D activities, which are internal sources of new products or new process innovation, and the practical content of new product and new process innovation of the industries.

6.2 Open Innovations of Fuel Cell SMEs

6.2.1 Contents and Channels of Open Innovation by YLJO

YLJO accumulated considerable technical skills in the control area focused on order production of automation facilities of door locking—in particular, the company could accumulate a considerable degree of technical skills and knowledge in the production process of user customized automation facilities. As the orders and requirements from customers gradually increased, the skills accumulated by the company were also extended to the area of one-touch fitting (related to fuel, plumbing automation) and could manufacture stack holder, fuel cells, etc. This company developed its main products through outside-in open innovation in the process of meeting with requirements from users or customer companies located on the front side of the value chain. But in this process, the company owner's long experience in electrical and electronic fields and aggressive intentions and efforts to obtain knowledge through specialized literature or external counseling acted as an important motive of open innovation. The CEO of this company, immediately sensing the necessity of continuous internalization of domestic and foreign advanced technical trends through systematic and continuous research collaboration with external specialized institutions, focused on new areas of fuel cells and raised the necessity of policy support for the relevant open innovation program.

The process of obtaining knowledge in the fuel cell area of this company (YLJO) is a representative case of outside-in open innovation from customers. More specifically, this company obtained considerable knowledge through informational exchange with university research teams in the process of producing. This inward open innovation is not fixed but shows diverse aspect of improvement and development of the technical skills in the fuel cell area of the company that are accumulated. In other words, although it was a one-sided relationship at the initial stage, the company was gradually able to produce final products through the processes of specification decision of ordered products, designing requested products, and finally the settlement of errors, after YLJO shared opinions and mutual discussions were conducted when an idea was suggested by the university research team. In other words, outside-in open innovation developed from simple order production to interactive studies. This company is strengthening its expectations and preparation for open innovation through production of new products in the fuel cell area and the creation of new markets by developing further skills and knowledge accumulated in this way through collaborative research.

6.2.2 Contents and Channels of Open Innovation by DSN

Second, in the case of DSN, various multidimensional activities and efforts for specific technological areas, i.e., outside-in open innovation for thermal spraying, are shown. In particular, this company is conducting more multidimensional outside-in open innovation through the accumulation of its own technical capabilities and improvement of internal research teams and research facilities. DSM is a company invested in by the Japanese and established as a company specialized in the thermal spraying business with POSCO (a large steel maker from South Korea) as its main customer. This company grew into the largest domestic company specialized in thermal spraying in the high heat area, additionally securing Hyundai Steel, Dongkuk Steel, Doosan Heavy Industries and Construction, KEPCO, and Hansol Paper as its main customer companies. DSM now possesses six global original patents in the thermal coating area as a result of expanding its research facilities and research teams.

By starting open innovation through the introduction of new technologies from Japan, DSM accumulated considerable technical skills in the thermal spraying area in the process of interacting with POSCO, its main customer company, and addressing its requirements. However, as the customer company grew into a global company, and as the technical requirements to DSM, the supplying company, increased, DSM was faced with the necessity to possess far more advanced skills than the customer company. Accordingly, DSM accumulated technical skills by using various approaches of outside-in open innovation such as (1) dramatically strengthening the capabilities of its internal research institute, (2) exploring ways of technical cooperation for benchmarking foreign technologies, (3) introducing advanced technologies, and 4) technical alliance. Specifically, in the case of advanced high-tech areas such as glass rolls and metal doughnut rolls, DSM decisively introduced technologies on a royalty basis and has continuously obtained technologies through a technical alliance with Stock of Germany about SOFC-related information. In addition, this company is making various efforts to improve its technical capabilities through cooperation with RIST in Pohang about ways of improving the thermal resistance and durability of thermal spraying related to separators. Further, it has been actively strengthening its efforts to secure technical capabilities through alliances with external research centers and industries such as through technical agreements of thermal spraying-related areas with TOCALO PTO of Japan since 1990. DSM is also continuously making efforts to understand recent technical trends by participating in various domestic and foreign conferences related to thermal spraying. Through these efforts, it has become the core company of cooperative research related to thermal spraying for domestic industries including POSCO. Consequently, DSM is being treated as an important partner in technical cooperation by big enterprises that have technical demand for thermal spraying in the fuel cell and solar energy area. It has already established a semiconductor-related thermal spraying company through spin-off that has become larger than the parent company.

Even in the case of DSM, which accumulated domestic top-class and world-class technologies in the thermal spraying area through active open innovation, this company has encountered difficulty in finding cooperative partners in the coating and surface treatment area because of the lack of an open innovation arena where the company can interact with internal or external industries with the same or different lines of business. In addition, this company is faced with considerable difficulties due to the lack of a systematic open innovation network with external industries or research institutes, although thermal spraying skills are considered to be strongly needed also in the solar energy area, not only in the fuel cell area.

6.2.3 Contents and Channels of Open Innovation by OT

The third case is open innovation of OT which is a medium-sized firm in solar energy. Starting as a production company of vacuum equipment for sputtering, OT is advancing into the area of production of solar energy (solar cell)-related products. This company is continuously securing new technologies and new knowledge through joint research with Kyungpook National University, Gumi Electronics and Information Technology Research Institute, and Nano Convergence Practical Application Center, and cooperative research with major research institutes outside of the region such as ETRI and Korea Electrotechnology Research Institute, and through continuous monitoring of customers' requirements and expectations. In addition, this company is in a strong position to secure new ideas and knowledge

Table 6.1 Real channels and contents of open innovation by fuel cell SMEs

Company	Concrete contents and channels of open innovation
YLJO	Meeting with requirements from users or customer companies located on the front side on the value chain Continuous research collaboration with external specialized institutions Informational exchange with university research teams
DSM	Introducing technologies on royalty basis and continuously obtaining technologies through technical alliance with Stock of Germany about SOFC-related information Alliances with external research centers and industries such as a technical agreement in the thermal spraying-related area with TOCALO PTO of Japan since 1990 A semiconductor-related thermal spraying company through spin-off and developed the company to become larger than the parent company
OT	Joint research with Kyungpook National University, Gumi Electronics and Information Technology Research Institute Continuous monitoring of customer firms' requirements and expectations Participating in academic societies (display and solar cell-related societies) Coping with customers' requirements and requests while producing customized products

Source: Yun and Mohan (2012) revised

through monitoring of foreign original patents. Further, this company is trying to understand the trends of market change and technologies through participation in academic societies (display and solar cell-related societies).

This company took the opportunity to advance into the solar energy industry by acquiring information and ideas from customer companies. OT supplied doping equipment to a company, and when OT saw the company apply the equipment to single crystal solar cells, it saw the chance to improve its technical capability related to solar cells in the process of connection with the company. In other words, this company accumulated technologies in the process of coping with customers' requirements and requests while producing customized products, and this allowed it to develop new products, a typical example of which is advancement into the solar cell industry.

We can summarize the real channels of open innovation by fuel cell SMEs. Most of all were the key open innovation channel (Table 6.1).

6.3 Open Innovation of SMEs in Medical Instruments and Intelligent Robot Industries

6.3.1 Contents and Channels of Open Innovation by OGV

First, OGV, an IT medical facility firm, focused attention toward the efforts for inside-out open innovation of S/W industries that develop wireless control systems with the customers of domestic and foreign smartphone end products industries. OGV's CEO and major research and production staff had previous experience of working for smartphone S/W industries targeting major domestic mobile phone enterprises. OGV is a company that the CEO and other major staff established by spinning out an existing company, and it is making efforts for various inside-out open innovation based on existing accumulated technologies as well as existing smartphone S/W technologies. In particular, in order to develop IT medical-related new products that applied and developed existing smartphone wireless control technologies, OGV is making efforts for various open innovation activities, not depending only on inside technical capabilities. First, this company is continuously securing information related to new technologies and knowledge through personal networks of its internal core researchers with research teams of their alma mater universities. Examples include cooperative tasks with domestic major mobile phone enterprises and cooperative research with regional and domestic major universities related with national research tasks of medical IT. In order to apply existing wireless control technologies to the IT-medical area, this company is conducting various cooperative activities such as establishing networks with domestic universities, customer companies, and relevant researchers to secure additional technical capabilities necessary for inside-out open innovation, and not just settling for internal technologies. Also, this company established another inside-out open innovation strategy of developing wireless control-applied S/W related to smartphones and is conducting various activities.

In a case analysis of this company, it was confirmed that the success of open innovation to apply its technologies to the IT-medical area is significantly associated with the level of open innovation with industries. Particularly in the case of IT-medical industries, which are newly emerging, it is considered necessary to arrange an opportunity for universities, companies, and relevant industries to practically grow at the same time through a broadscale extension of large-scale open-type original research programs focused on universities.

6.3.2 Contents and Channels of Open Innovation by DTS

Second, in the case of DTS, which is a firm in the IT-medical device parts industry, the direction and contents of nonuser inside-out open innovation of this industry can be examined. As in the research of Hippel (2005), medical devices make up a representative area where various innovations occur through user innovation (in this case, by doctors). Most medical device user innovation by doctors does not just remain at user innovation but is also connected to the producer and further to open innovation. In the case of DTS, the CEO and executive director of the company are professional managers who worked for major domestic medical device companies for a long time, and the director of the research center and relevant professionals were scouted from implant-related companies at the time of company establishment or after the establishment. Despite this composition of manpower, the business staff of this company is frequently receiving user innovation requests from doctors and connecting them to the development of new products, and they are also systematically collecting user innovation requests suggested from doctors participating in implant seminars, which they are using as sources for product innovation. Examples include drills that do not touch nerve cells and an innovated product that can adjust denture angles.

This company is systematically collecting user innovation ideas from doctors but also has a system of verifying in advance the merchandising of innovation ideas according to market prediction and the management strategy of company directors. DTS is, in fact, exposing its limitations of not conducting lively user-based open innovation because innovation ideas of users sometimes are not sufficiently connected to the innovation of new products through the process of prior verification of the management. DTS is indirectly actualizing user innovation such as utilizing doctors as a consultation group but has a limitation in user-based open innovation of the implant itself. This company is trying to obtain new knowledge and technologies from outside through various seminars and meetings by opening its seminar rooms, which is considered as a result of the CEO's judgment and volition that user innovation and aggressive acquisition of external ideas occupy an important place in the innovation of industries.

But this company is actively making efforts to advance into new areas such as medical devices of new areas, medical devices for operation, and new devices for the dental area, through purchase of technologies from relevant industries, establishment of spin-off companies through joint investment with external research institutes, and through M&A with external industries. In other words, this company is actively making efforts for open innovation by way of technology licensing, joint investment, and spin-off.

6.3.3 Contents and Channels of Open Innovation by YJM

Third, through a case analysis of YJM, the limitations and possibility of open innovation of new growth engine industries with important customers were identified. YJM is a spin-off company from Yujin Robot with industrial robot technology in 2002. Yujin Robot, YJM's parent company, was established in March 1988 as a company specializing in intelligent service robots (nursing and support robots for the elderly, patients, and the disabled, family robots, communication robots, cleaning robots, service robots), with capital of KRW 8.247 billion and about 100 employees, and its major products are cleaning robots, ubiquitous home robots, entertainment robots, and toys. YJM was established with the initial technical background of Yujin Robot and is mainly engaged in the steel milling area by targeting a niche market among industrial robots. This company presents a representative case of accumulating technologies mainly in the process of coping with requirements and expectations from an important customer, POSCO. In the first stage, it dealt with a simple order production, but as this company settled technical problems of POSCO and as its technologies were accumulated, their relationship developed to the stage of joint research. Further, this company shows a considerable effect of technology accumulation by independently applying for 50% of the patents established in the process of its relationship with POSCO and jointly applying for the remaining patents with POSCO.

As for the process of technology accumulation, POSCO tried to solve some problems of the products that it had been developing for a few years with trial and error in its research center by placing orders with YJM, which successfully settled the problems and made a partnership with POSCO. Currently, this company has reached the stage where it can discuss ideas with POSCO face to face, and this company is developing products when it receives organized requirements from POSCO. In addition, this company is trying to acquire various knowledge and technologies from outside through national research projects, about per year, and making efforts for acquisition of various knowledge and technologies through joint research development with DMI, Kyungpook National University, and Yeungnam University in the region. This company is also obtaining various knowledge and technologies in the robot area through online information and user communities, and it is systematically grasping the worldwide robot-related technical trends through participation of its researchers in academic societies and conferences.

SMEs in medical instruments and robot industries carried out open innovation intensively in such channels as research projects with university and national labs rather than user firms in the supply chain, as seen in Table 6.2.

Table 6.2 Real channels and contents of open innovation by SMEs in medical instrument and robot industries

Company	Relevant channels of open innovation
OGV	OGV is a company that the CEO and other major staff established by spinning out an existing company Cooperative tasks with domestic major mobile phone enterprises and cooperative research with regional and domestic major universities Conducting various cooperative activities such as establishing networks with domestic universities, customer companies, and relevant researchers to secure additional technical capabilities
DTS	The CEO and executive director of the company are professional managers who worked for major domestic medical device companies for a long time Utilizing doctors as a consultation group New knowledge and technologies from outside through various seminars and meetings by opening its seminar rooms Technology licensing, joint investment , and spin-offs
YJM	Accumulated its technologies mainly in the process of coping with requirements and expectations from an important customer, POSCO Trying to acquire various knowledge and technologies from outside through national research projects Making efforts for acquisition of various knowledge and technologies through joint research development with DMI, Kyungpook National University, and Yeungnam University in the region Obtaining various knowledge and technologies of the robot area through online information and user communities

Source: Yun and Mohan (2012) revised

6.4 Closed Innovation of SMEs

6.4.1 Cases of Closed Open Innovation of Large-Scale Process Industries

STL advanced into the solar energy industry when "Lucky Material" entered into a technical partnership with SILTEC of the USA for production of silicon wafers. After changing its company name to STL in 1990, this company took over the silicon wafer business of "Lucky Material" in 1991. This company had a solar cell line in the 1990s (1996, 1997, 1998), and with some slow down during the Asian financial crisis, it has been actively conducting research on solar energy since 2004. This company has no separate solar cell development team, but the existing technical development team gave some support to research related to solar cells. STL had no customers in the group, but when the group started solar energy business, it emerged as a new customer of this company. LG Chemistry, which produces polysilicon; STL, which produces wafers; and LG Electronics, which produces end product such as solar cells, were established as front and rear connected industries.

STL started its business with semiconductor wafers. It is considered that this company is establishing a closed-type research system inside the company without any cooperative research with the outside because semiconductor wafers are a hi-tech industry, whereas solar energy industry has a low entry barrier, and existing semiconductor technologies are sufficient to conduct research on solar energy. Currently, this company has almost no external interaction (channel) for technical development of solar energy. But it had experience of interacting with universities through the forms of internship or small tasks (industry funding: Kumoh National Institute of Technology, Yeungnam University, and Kyungpook National University). In some low-efficiency research topics, this company sometimes gives tasks to universities, which show more results. STL secures its raw materials mainly from Japanese companies and devices from neighboring Daegu-Gyeongbuk companies. The solar cell industry is in need of a consortium. But this company does not require a consortium because of its internal connections, and this company can independently produce substrates, and also it does not particularly consider interaction with universities. However, in order to lead the world market and to secure future competitiveness, a large-scale open-type original research program is necessary for research development activities such as the development of mass storage ingots or production of premium cells with more than 20% efficiency. The lack of open-type research in a large-scale process industry to lead future industries is closely linked to the market failure of the industry. Accordingly, the government should actualize joint research with universities including big enterprises and medium and small firms by developing future original open-type research programs in solar energy areas.

6.4.2 Risks of Self R&D Under Closed Innovation System

A self R&D case under closed innovation of "A" company explicitly suggests the risks that closed innovation brings to industries. The major product of this company is car motor automation facilities, which this company supplied mainly to second vendor industries. This company invested most of the profit obtained through this area of business to relevant internal R&D. Through this R&D, this company has 12 patents including original patents, 8 patents on new devices, 1 program registration, 1 international patent pending, 7 patents pending, and many other original technologies not yet applied for patent.

The CEO of this company majored in electronic engineering in university and further studied fuzzy electrical engineering in graduate school, entered an F-16 fighter-related company, and established his current company after resigning from the previous company. His sources of ideas for new technology development are Hannover Exhibition, Germany and research papers in relevant areas. He explained that he made efforts to develop technologies that could realize his belief that "technology can be called technology only if it can be universally and publicly enjoyed by humankind, with its economic value." But he never attempted any cooperation with industries in the same line of business, relevant industries, universities, and national research institutes. Consequently, this company is faced with a situation that is unbearable for his company due to time and expenses involved in the process of commercialization of developed technologies.

The CEO of this company developed original patents related to generation by using motors and tires and generation by using vibration, based on his basic knowledge related to motor automation facilities and knowledge accumulated in the process of development and production of many products combined with his personal intellectual curiosity. Among them, there are many prospective patents, but products beyond existing products are not realized in the market, which is a problem. The new technologies of this company have the limitation that the market is not understood due to the closed system of the company and that they do not reflect customers' requirements. Furthermore, the quality of new products is not guaranteed due to the lack of agreement of the researcher group, and this company is in a situation where it cannot ask for demands and agreement from relevant industries.

6.4.3 Closed Innovation of Local Subsidiary Company of Original Technology-Based Industry

The closed innovation case of a Korean subsidiary company of MG is a typical case that shows that even a high-tech industry can face considerable danger when its local subsidiary is insensitive to open innovation. MG Korea is a Korean subsidiary company of MG Crucible established in the UK in 1856, and its main products are thermal ceramics, carbon brushes, engineered carbon, and graphite heat exchangers. This company has global top technology in thermal ceramics.

MG Korea had a considerably large-scale domestic market, but it was considerably reduced in terms of current total sales amount of about KRW 30 billion, with its total employees about 150 and 5 research staff. MG Korea is involved in cooperation, in a considerably passive way, with domestic industries and research institutes that requested cooperation, depending only on the high technical capability of its parent company rather than independent and voluntary activities for technical development. MG Korea accumulates its technologies mainly from internal R&D tasks, and it depends on its parent company for most of its technologies. The parent company partly recognizes the necessity of open innovation, but it is sought mainly in its relationship with China. Due to the original technology of MG, tasks of joint research of the ceramic area related to solar cells are being conducted with Samsung Advanced Institute of Technology and Korea Electronics Technology Institute, but also in a very passive way. This company presents a typical case of closed innovation of a local subsidiary company directly invested in by technology-based foreign companies. This company can secure additional markets and develop new markets in the solar cell area only through actualization of cooperative research with domestic industries. In other words, it could be confirmed in this case that development of local markets and a research foundation is necessary for actualization of open innovation of foreign direct investment industries.

As seen in Table 6.3, three closed innovation example firms had high technology and high potential, but these factors nonetheless could not exert a positive effect on the development of these firms. Closed innovation causes a disconnection between technologies and markets in these SMEs. Closed innovation in SMEs causes firms to endure more serious situations than big businesses because they cannot survive in the present circumstances due to the disconnection between the technology and the market.

Table 6.3 Realities of closed innovation of SMEs

Company	Characteristics
STL	Establishing closed-type research system inside the company without any cooperative research with outside The lack of open-type research of large-scale process industry to lead future industries is closely linked to the market failure of the industry
"A" company	Through relevant internal R&D, this company obtained 12 patents including original patents, 8 patents on new devices, 1 program registration, 1 international patent pending, 7 patents pending, and many other original technologies not yet applied for patent Does not understand the market due to the closed system of the company
MG Korea	Depending only on high technical capability of its parent company rather than independent and voluntary activities for technical development Technologies mainly from internal R&D tasks, and the firm depends on its parent company for most of its technologies

Source: Yun and Mohan (2012) revised

Research Question

1. Select any firm near you, and examine and find contents and channels of open innovation from it.
2. Find, examine, and analyze any firm that provides an example of closed innovation.

References

Chesbrough H (2003) Open innovation: the new imperative for creating and profiting from technology. Harvard University Press, Cambridge, MA. pp 1–19, 43, 61, 93–112, 113–133

Chesbrough H (2006a) Open innovation: a new paradigm for understanding industrial innovation. In: Chesbrough H, Vanhaverbeke W, West J (eds) Open innovation: researching a new paradigm. Oxford University Press, Oxford, pp 3–11

Chesbrough H (2006b) Open business models: how to thrive in the new innovation landscape. Harvard Business School Press, Boston. pp 15, 20, 196–203

Kim WC, Mauborgne R (2004) Blue ocean strategy. Harv Bus Rev 82–10:76–84

Laursen K, Salter A (2006) Open for innovation: the role of openness in explaining innovation performance among UK manufacturing firms. Strateg Manag J 27(2):131–150

Von Hippel EA (2005) Democratizing innovation. MIT Press, Cambridge, MA

Yun J-HJ (2008) Analytical research on innovation networks and openness of regional clusters: focused on Sungsuh and Gumi industrial complex. Science & Technology Policy Institute, Seoul

Yun JHJ (2009) Geographical boundary of open innovation: sources within and beyond cluster. Paper presented at Atlanta conference of science and technology, October 2–3

Yun JJ (2015) How do we conquer the growth limits of capitalism? Schumpeterian dynamics of open innovation. J Open Innov Technol Market Complex 1(1):1–20

Yun JHJ, Mohan AV (2012) Exploring open innovation approaches adopted by small and medium firms in emerging/growth industries: case studies from Daegu-Gyeongbuk region of South Korea. Int J Technol Policy Manag 12(1):1–19

Yun JJ, Jeong E, Yang J (2015) Open innovation of knowledge cities. J Open Innov Technol Market Complex 1(1):1–20

Dynamics of Open Innovation

7

Abstract

We created conceptual models that might be used to analyze and forecast the dynamic effects of open innovation. We then applied these models to the smartphone sector using a model-based analysis approach. In addition, we built an open innovation simulation model for the smartphone sector. The dynamic model of open innovation will link logic and concepts relating open innovation, complex adaptive systems, and evolutionary change. The model was used to analyze the dynamic effects of open innovation strategies and open innovation simulation for the selection of future strategies.

Keywords

Open innovation • Complex adaptive system • Evolutionary change • Smartphone

7.1 Introduction

As the knowledge-based economy develops, the amount of knowledge in the world is rapidly increasing along with the velocity of circulation. Firms are increasingly utilizing not only their own technologies but also external knowledge and technologies. In addition, the open innovation phenomenon is rapidly spreading into industry, nationwide, and worldwide, as firms provide their unused technologies to be utilized by others. User innovation, customer innovation, collective intelligence, crowd sourcing, and open source innovations will be open innovation (OI) in that they are innovation based on transfers across the boundaries of knowledge and

This chapter is mainly based on the following paper.
Yun J.H.J., Won D.G., Park K.B. (2016), Dynamics from Open Innovation to Evolutionary Change. *Journal of Open Innovation: Technology, Market, and Complexity.* 2(7), 1–22.

J.J. Yun, *Business Model Design Compass*, Management for Professionals,
DOI 10.1007/978-981-10-4128-0_7

technology. The life cycles of cutting-edge products are becoming shorter and shorter, and brand new products of a firm are routinely being imitated by others. This is called the commodity trap and is increasingly common. Consequently, as a process enabling the relentless innovation of technology, open innovation is receiving more and more attention (D'Aveni 2010). Given this situation, we wanted to answer the following questions:

1. What kind of dynamic effects for business firms can result from complex innovation systems and market evolution driven by open innovation strategies and open business models?
2. What kind of effects can open innovation at the firm level give to and take from complex adaptive systems such as the national innovation system (NIS), regional innovation system (RIS), and sectorial innovation system (SIS) (Nelson and Winter 1982)?
3. How can firms escape falling into the commodity trap and suffering from a harmfully shortened product life cycle when engaged in the dynamic process of open innovation?
4. How do dominant design and technological regime appear, change, and disappear in the dynamic process of open innovation?
5. How are specific technologies or other knowledge selected by firms in the market during the dynamic process of open innovation?

We seek to establish a theory about the whole process by which open innovation is realized at the level of business firm. Speaking concretely, we seek a theory about all the processes by which new ideas or technologies are adopted by a firm, how they are used to create new products or processes, and how, in the end, they are incorporated into dominant design. Entire fields of industry are increasingly confronted by the perils of the commodity trap, in which imitation or pursuit of cutting-edge products is made within very short time frames. For this reason, a firm needs to dynamically analyze the impacts of its own open innovation strategy on the introductory stages of new knowledge, technologies, or ideas. In analyzing open innovation of a firm, we cannot understand and analyze fully the whole process of open innovation without analyzing the dynamic process of specific open innovation strategies. First of all, concrete open innovation strategies of firms, and analysis of the dynamic processes involved, are more important than ever. Open innovation at the firm level is no longer an option but rather a must for the survival of not only corporate giants like IBM, 3 M, and Intel but also small- and medium-sized enterprises (SMEs) as well (Yun and Mohan 2012b).

The requirement of a new approach for firms to deal with the increasing open innovation phenomenon in the form of open innovation strategies, business models, user innovation, collective intelligence, and crowd sourcing is increasing. Firms need new ways to escape the commodity trap and to prevent injury from short product life cycles.

There is also a need for connections between open innovation at the firm level; complex adaptive systems such as regional innovation system (RIS), sectorial innovation system (SIS), and national innovation system (NIS); and evolutionary change in markets. There needs to be a research framework aimed at solving this problem. Finally, we want to understand the total cycle of innovation in firms: from new ideas to new products and from dominant design to choice of technological regime (Lee and Lim 2001). There are already several theories intended to answer these questions as follows (Heredero and Berzosa 2012).

First, resource- and knowledge-based theory treats open innovation as a way to exploit resources and knowledge complementarities (Mowery et al. 1996; Das and Teng 2000; Nonaka 1994; Simon 1991). This resource-based perspective focuses on strategies for exploiting existing firm-specific assets (Teece et al. 1997). Well-known companies like IBM, Texas Instruments, Philips, and others appear to have followed a "resource-based strategy" of accumulating valuable technology assets, often guarded by an aggressive intellectual property policy (Teece et al. 1997). If control over scarce resources is essential for profit, such issues as skill acquisition, the management of knowledge and know-how, and learning become fundamental strategic issues (Shuen 1994). However, this theory cannot explain the dynamic changes that originate from the firm, such as the commodity trap and shortened product life cycles.

Second, according to the transaction cost theory, open innovation will decrease transaction costs by vertical disintegration of firms. This theory was derived from the coarse theorem and new institutional economics (Coase 1937; Williamson 1991; Kogut 1988). On the other hand, dynamic capabilities and transaction cost coevolve according to examples such as the mortgage banking industry in the USA, which showed a shift from integrated to disintegrated production, and the Swiss watch manufacturing industry, which went from disintegration to integration (Jacobides and Winter 2005). Transaction cost theory can explain the usefulness of open innovation in restricted areas such as cost reduction. From the viewpoint of focusing on the coevolution of transaction cost and dynamic capability, a systematic structure with new knowledge and technology should be extended into a firm for reduction of transaction cost, and then coevolution will occur (Jacobides and Winter 2005). Although this logic does not concentrate on the strategy of a firm, it coincides with the direction of this chapter in that it focuses on establishing and analyzing a model of the dynamic process of open innovation.

Third, the dynamic capability theory is a history-friendly model at the economic level, which can explain changes in the economy from the introduction of new technology or knowledge by firms. This theory is based on Schumpeterian economics (Malerba et al. 1999a, 2001, 2008; Nelson and Winter 1982; Yoon and Lee 2009). This model can be applied to a simulation model by the history replication method and can then be used to predict and analyze dynamic changes in economic phenomena through a history divergent simulation (Malerba et al. 2001). Furthermore, this theory analyzes the process of dynamic change (macroeconomic effect from microeconomic phenomena) through the simulation method. This is similar to the case of competition and industrial policies in a "history-friendly" model of the evolution of the computer industry. Namely, the basic analysis target of the dynamic capability theory is not a firm but an economic phenomenon. Consequently, this theory has limits for concrete analysis of the process of dynamic change caused by changes at a firm (e.g., business model or strategy). After all, the analysis beyond history replication is left in a black box, because this theory has adopted an approach based on simulation. This chapter focuses on the analysis of the black box itself, the dynamic change which open innovation brings to a firm.

Fourth, again according to the dynamic capability theory, collaborative innovation is established to develop the dynamic capabilities of a firm and thus enhances its competitive advantage. This theory was proposed and developed by several firm strategy research groups and Schumpeterian economists (Teece and Pisano 1994; Teece et al. 1994; Eisenhardt & Martin 2000; Zollo and Winter 2002; Helfat et al. 2007; Teece et al. 1997; Arthur 1994; Chesbrough and Teece 2002). The dynamic capabilities framework analyzes the sources and methods of wealth creation and capture by private enterprise firms operating in environments of rapid technological change (Teece et al. 1997). According to this theory, the competitive advantage of firms is seen as resting on distinctive processes (e.g., ways of coordinating and combining) shaped by the specific asset positions of each firm (e.g., portfolio of difficult-to-trade knowledge and complementary assets) and the evolutionary paths it adopted or inherited (Teece et al. 1997). The dynamic capability theory forcefully explains dynamic aspects with which a firm is faced (e.g., having to secure knowledge or technology, managing corporate cooperation, and path dependency resulting from acquisition of new knowledge assets). This theory also sets corporate strategic targets to maintain a firm's competitiveness by keeping rival firms from imitating and replicating their own creative products. However, it does not include direct mention of concrete corporate strategy aimed at preventing the elements comprising dynamic capability from being imitated and replicated. Dynamic capability theory cannot directly explain the trigger of dynamic capability. There is no sufficient explanation of the starting point of the introduction of new ideas, knowledge, or technology as a dynamic activity performed by a firm. To discuss dynamic capability at a corporate level, this chapter seeks to build up a model of the dynamic processes involved in open innovation to analyze those processes starting from a decision by a firm to adopt an open innovation strategy and then apply it to the current smartphone sector.

Fifth, evolutionary theories of business activity note that some firms struggle to meet the demands of their environments and reside at the margins of survival (Fortune and Mitchell 2012). In turn, selection processes remove struggling firms from the business landscape, if they fail to improve (Nelson and Winter 1982; Aldrich 1999). According to this theory, firms survive by overcoming the obstacles of dissolving their obsolete skills or assets and of acquiring the new skills or assets required. For example, Cisco personnel suggested that the company used acquisitions to overcome market failure in the discrete exchange of organizational capabilities as it sought to upgrade its technical and market resources (Fortune and Mitchell 2012). This perspective is significant as a framework to understand economic phenomena based on the behavior and strategy of firms. At the same time, this theory does not fully explain both radical innovation, which arises frequently and appears in unpredictable ways, and acquisition and dissolution aimed directly at knowledge and technology in the knowledge-based age. In other words, what is required is a direct analysis of concrete dynamic processes at a corporate level. Thus, this chapter seeks to identify how evolutionary technology management and strategy, i.e., open innovation technology management strategy at the corporate level evolves in a complex adaptive system.

For these reasons, we need to develop a theoretical concept model that can explain the processes from open innovation of new ideas or technology, to the appearance of dominant design, and the evolution of national innovation system, RIS, or SIS. The growing body of literature on strategic alliances, the virtual corporation, buyer-supplier relations, and technology collaboration indicate the importance of external integration and sourcing (Teece et al. 1997). Namely, there is a growing necessity to analyze the dynamic process of open innovation, and that is the subject of this study.

7.2 Model Building

We build up a model that connects open innovation, through complex adaptive systems, with evolutionary change (the OCE Model, see Fig. 7.1). There is a deep relationship between these three factors (open innovation, complex adaptive systems, and evolutionary change), and they are arranged in conceptual order in the name of the model, not in temporal order. Conceptually, open innovation at a firm goes through a complex adaptive system and then leads to evolutionary change. However, in reality, a specific complex adaptive system can trigger open innovation through evolutionary properties at any given firm. The OCE model in Fig. 7.1 is based on the conceptual order needed to analyze the dynamic changes triggered by open innovation at an individual firm (Yun 2015).

The basic subject of open innovation is an agent (in this case, a firm). The agent chooses actions based on its independent judgment. Its actions influence other agencies or environments, and it can also be influenced by them. In that sense, social organizations, individuals, and government can be the subjects of open innovation as well. Firms make various degrees of innovation (incremental to radical, inbound, or outbound) through diverse channels (corporate open innovation influences national innovation system, RIS, and SIS). This complex adaptive system exhibits various levels of emergence. At the level of price differentiation or product differentiation, examples might include change of dominant design, creation of new firms, or even emergence of new sectors. This complex adaptive system influences corporate open innovation by way of strange triggers based on client features. These might include degree of suitability of fascination of customers with the technology regime; technological capabilities of related SIS, RIS, or national innovation system; and the existence and level of rival firms or suppliers. Of course, the unique historical heritage, location, ecosystem, or environment of the innovation system are unique features that work as a strange trigger with certain effects on the diverse open innovation activities of a firm.

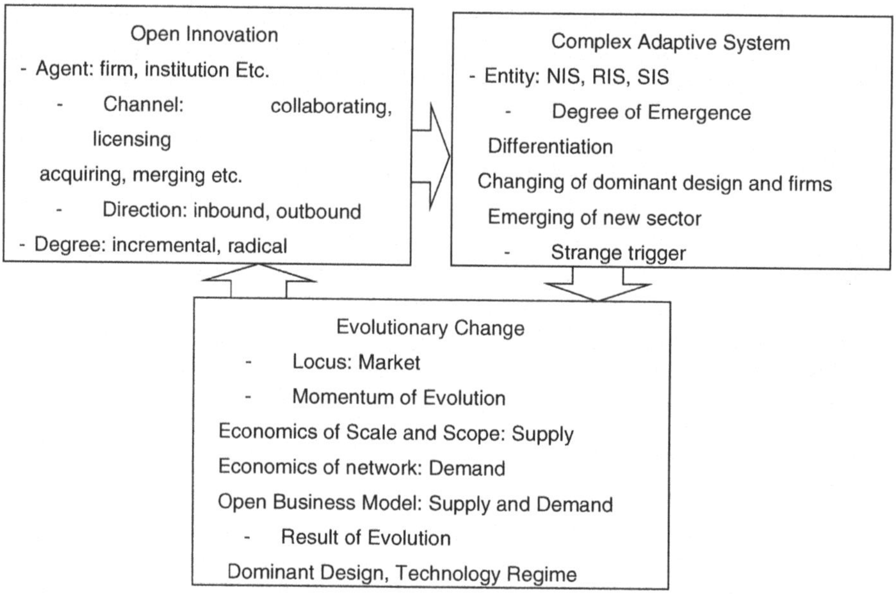

Fig. 7.1 OCE model concept

Corporate open innovation goes through evolutionary stages in the market, blooming into various types and levels of emergence or being influenced by strange triggers, under complex adaptive systems. The basic locus of evolution is a market. Corporate open innovation shows up as dominant design thanks to various evolutionary factors (e.g., economies of scale and scope, economics of networks, and open business models). After all, corporate open innovation creates market lock-in, by initiating path dependence and forming the technology regime.

The degree of corporate open innovation creates a variety of evolutionary effects according to the degree of complexity of the complex adaptive system.

In the case of A in Fig. 7.2, the degree of corporate open innovation is high, and the degree of complexity of the complex adaptive system (CAS) is high as well. The degree of complexity of the CAS is high and has two meanings. One is that the openness to a new product is great, because customer fascination with such products is high. The other is that the production and distribution of new knowledge and technology are brisk, because the capacity for research and development (R&D), as well as the technology of various agencies in the system, is great. In other words, the openness of the innovation system itself is great, because the strange trigger of the CAS is robust. In the case A firm in Fig. 7.2, a firm fully commits to global open innovation, which brings about continuous evolutionary change.

In the case of B in Fig. 7.2, though the degree of corporate open innovation is high, the degree of complexity in the CAS is not as great. Here, there are considerable difficulties in connecting the fruit of corporate open innovation with evolutionary change. In this case, firms are gradually confronted with limits to performing open innovation activities at a high level of energy. Here also the fruit born by the

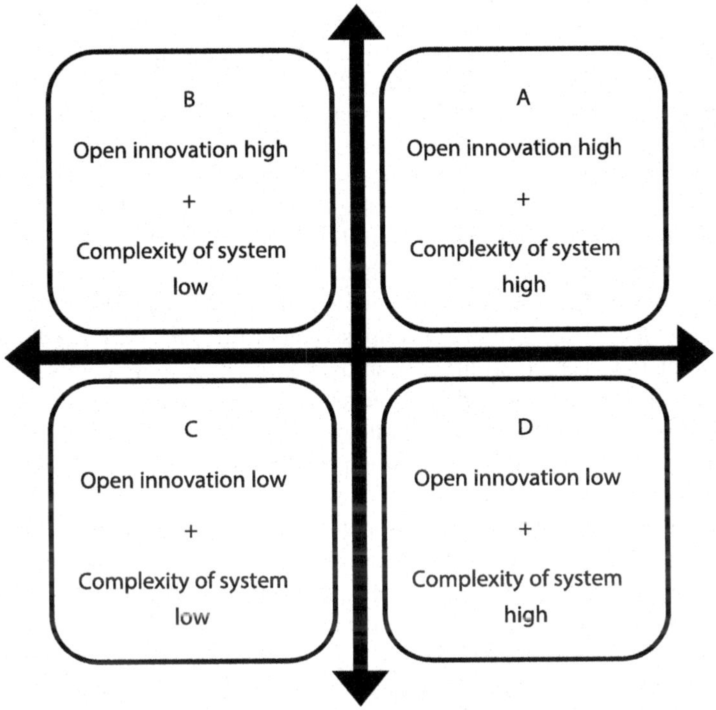

Fig. 7.2 The relationship between open innovation and complexity of system

open innovation activities of the firm are neither acknowledged by the customers of the innovation system itself nor supported by technological capability, and the firm faces difficulties owing to the discord between the products of the firm and its market. In this regard, a firm needs to develop an open innovation system appropriate for it, and concentrate on it, or it needs to adapt its own open innovation for the CAS in which it is involved.

In the case of C in Fig. 7.2, not only the level of corporate open innovation but also the degree of openness of the CAS is low. There is little reason to hope for enhancement of the corporate open innovation and openness of the innovation system, without external stimulus or institutional change. In this case, a corporate strategy aimed at enhancing the degree of open innovation should be developed and political measures taken at the same time to increase the openness of the CAS.

Specifically, R&D programs should be developed to directly improve corporate open innovation, to establish cooperative research with university and national research institutes, and to promote the employment of excellent research personnel. Further, at the level of national innovation, it is also necessary to introduce various technologies and to invite excellent R&D personnel from abroad, to promote participation in global R&D programs, and to concentrate on nurturing domestic R&D personnel through attendance at famous foreign universities with a focus on research. In addition, at the level of sectorial and regional innovation, it is necessary to prepare political measures to enable improvements in system openness and complexity.

The situation indicated by D in Fig. 7.2 is typical of most current, cutting-edge enterprises, corporate giants, and technology-based small and medium-sized enterprises (SMEs). The openness of the innovation system is low, and the degree of its complexity is high. For this reason, new corporate products fall into the commodity trap so that the life cycles of new technology are already very short or they face the danger that the technology life cycle will be shortened. Such firms have no option other than to keep creating new knowledge, technologies, or ideas, through continuous, active open innovation. As the openness of an innovation system is low and the degree of its complexity is high, it is impossible for firms to make technology innovation sufficient to keep rival firms from overtaking them or to completely protect their own technologies with patents. Firms have no choice but to form corporate organizations, to develop corporate strategies, to build a corporate production system aimed at continuous open innovation, and to build open innovation into the entire product life cycle.

As mentioned above, open innovation creates evolutionary change through corporate activity and coevolution with the CAS. Corporate open innovation activities influence the innovation system itself and, at the same time, are influenced by the innovation system, which brings about resonance and coevolution with other firms influenced by the innovation system.

The OCE model can be proved and analyzed by using documents, case studies, surveys, and social experiments (e.g., ultimatum game or iterated prisoner dilemma). Further, the OCE agent-based model (ABM) can be built to simulate real situations. Normally, agent-based models occasion the problem of validity inevitably (Carcia et al. 2007). Such approaches as a conjoin analysis have traditionally been used to secure the validity of agent-based marketing models.

Using the OCE model, more accurate analyses can be made of the long tail phenomena gradually increasing in online and mobile markets (e.g., e-books and music). It is possible not only to make a direct analysis and explanation of open innovation of firms related to crowd sourcing, or of the gradual increase in collective intelligence in diverse fields (e.g., Wikipedia, Quirky, Threadless), but also the increasing profits occurring at App Store and the like. The OCE model is also very useful for analyzing major fields in which open innovation of firms is occurring (e.g., smartphones, e-books, online music, pharmaceuticals, and consumer electronics). Open innovation in all the industries is gradually being strengthened thanks to continuous development of the knowledge-based economy. For this reason, it is expected that industries or sectors for which open innovation analysis of firms using the OCE model is intended will continue to increase in the future.

We will now assemble the OCE model in three stages. First, we will build up open innovation factors, processes, and their connections with complex adaptive systems (Yun and Mohan 2012a). Second, we will build up diverse complex adaptive system factors and their relationships with open innovation and evolutionary change (Yoon and Lee 2009; Yun and Mohan 2012b). Third, we will construct evolutionary change resulting from the complex adaptive system and its interaction with open innovation (Malerba et al. 1999a, Malerba et al. 2008).

7.3 Construction of OCE Model

7.3.1 Open Innovation in the OCE Model

The conceptual boundaries of open innovation, as the target of the OCE model analysis, are based upon open innovation, user innovation, collective intelligence, crowd sourcing, and the open source approach in software development. All innovation based on the inflow and outflow of technologies, knowledge, and ideas crossing the boundary of firms are considered "open innovation" and the intended target of the OCE model.

On one hand, the concept of an agent of open innovation has undergone substantial change over time. Schumpeter thought of an entrepreneur, a person, as the agency of innovation in the initial stage of his research and a large company as the agency in the latter stage (Schumpeter 1934, 1942). After discussion of the strategy at corporate or national level became active with Porter (1980, 1990), actual approaches to open innovation strategy at institutes such as firms, social organizations, and government agencies have been discussed, analyzed, and carried forward diversely in direct or indirect methods. In the discussion centered on the firm as the basic agent in OCE model analysis, open innovation inside the firm becomes the target of its strategy. On the other hand, open innovation outside a firm, which can result in market or system failure, is the target of government policy.

The OCE model, as an open innovation channel, takes into consideration the factors on the technology push side, as well as the factors on market-driven side (e.g., collaboration, acquisition, merging, licensing, customers, suppliers, competing firms, universities, and national research labs). In reality, an apparent disjuncture between changes in technology and productivity can be observed, for instance, during the so-called productivity paradox of the 1980s and 1990s (conditions far from unusual in history—Tunzelmann and Wang 2007). The reason is that, first of all, in shaping production function, traditional theory or dynamics capability theory failed to take into consideration the source of new knowledge and technology. Not only capabilities of producers but also capabilities of consumers should be taken into consideration as bases of productivity or dynamic change (Tunzelmann and Wang 2007). One of the important considerations in the OCE model is the various channels through which knowledge and technology flow in. For example, open innovation considers merge and acquisition (M&A) as an open innovation channel to acquire tacit knowledge. The embedment of key capabilities and the knowledge that they embody often motivate firms to acquire an entire entity to obtain these capabilities, as opposed to simply licensing specific goods or hiring employees (Capron et al. 1998). In fact, the higher up the corporate scale a firm is, the more it seeks complete, open innovation, including tacit knowledge as well as codified knowledge through merge and acquisition (M&A).

From the corporate viewpoint, the degree of open innovation varies from incremental, representing the improvement of existing products, to radical innovation, representing the launching of completely new products on global markets as well as into domestic markets. The difference in the degree of open innovation causes the difference of emergence at various levels of the complex adaptive system (CAS). Regarding the relation between the level of open innovation and corporate achievement, there is a reverse U-curve in a quantitative analysis of the relation between the open innovation and corporate achievements of many firms (Laursen and Salter 2006). However, the relation between the degree of corporate open innovation activity and corporate achievement will vary according to the corporate environment. According to environmental dynamism or heterogeneity, the effects that the ordinary and dynamic capabilities have on the relative performance of firms were different (Drnevich and Kriauciunas 2011). Namely, it is useless to make a quantitative analysis of the relation between the degree of corporate open innovation and achievement. Rather, it is reasonable to make an analysis of the dynamic process of the corporate open innovation strategy, that is, the process for achieving the open innovation strategy of a specific firm, which is what the OCE model analysis is intended for.

A quantitative analysis of the determinants of open innovation showed a variety of factors that determine the success of open innovation such as the attitude toward openness, entrepreneurship, internal system for openness, and capability for corporate absorption (Yun and Mohan 2012a). The RIS, national innovation system, or SIS under which a firm functions (as factors external to the firm) were presented by way of a quantitative analysis of factors to determine the level of achievement of open innovation. According to an OCE model approach, related to analysis of the processes of dynamic change resulting from a corporate open innovation strategy, corporate achievement is not determined by the specific factors of open innovation. Rather, the factors that determine corporate achievement during the dynamic processes of open innovation work differently, and in some cases, the same factors have quite different effects. Namely, it is possible to find the concrete factors that influence the achievement of open innovation at the corporate level, and to analyze the influence, only through an analysis of the dynamic effects of open innovation.

On one hand, the degree of corporate open innovation determines how well the firm catches up with the leading firm in the belonging sector. Catch-up strategies basically assume three patterns such as pass-following catch-up pattern, pass-skipping catch-up pattern, and pass-creating catch-up pattern, which have target sectors of other national innovations (Lee and Lim 2001). In a knowledge-based economy, when the technological life cycle is being shortened, technological catch-up types move from a pass-following catch-up pattern based on a closed innovation strategy to a pass-skipping catch-up pattern pursuing a medium degree of open innovation, and a pass-creating catch-up pattern with a high degree of open innovation. The higher the degree of corporate open innovation, the more rapidly and creatively a related firm can follow up. The model to determine a catch-up pattern includes, as internal determinants, factors representing open innovation, such as access to an external knowledge base or other available knowledge and resources.

7.3.2 Complex Adaptive Systems in the OCE Model

Currently, the topic of complexity is attracting a great deal of interest, but there remains a question of what can be said meaningfully about complexity (Simon 1995). There are several complexity theories (e.g., mathematical theory of chaos) that deal with the complexity of nonlinear dynamic systems, whose long-term behavior is unpredictable. Systems theory, about systems that possess many interacting components, deals with another form of complexity. There is also a theory of computational complexity, which uses agent-based modeling and is applied to physical and economic issues all together. Complex systems arise naturally in the economy because economic agents, whether they are banks, consumers, firms, or investors, continually adjust their market moves, buying decisions, prices, and forecasts in response to the situations these moves, decisions, prices, or forecasts together create (Arthur 1999, 2009). The complexity in this study includes computational complexity as it uses a computer for analysis. The factors used to configure the system respond to, and have an influence on, the system. For this reason, it is called a complex adaptive system. Enterprises, which are representative agents and which make up complex systems, are not just collections of production factors, they are "repositories of competence" that create, coordinate, and deploy knowledge. In this case, it is the knowledge of the "specific connections that seem to work in a particular environment" (Potts 2001).

Complex adaptive systems enhance creativity of firms and prevent them from falling into the commodity trap or at least retards the shortening of the life cycles of their products. In this sense, complex adaptive systems have a very positive value. In other words, that the complexity of an innovation system is great means that new knowledge and technology flow into the innovation system continuously (Won et al. 2015), increasing not only the amount of knowledge but also the speed of knowledge distribution, in the system. Consequently, the determinant of complexity is, first of all, the degree of open innovation that an individual firm forms in the complex adaptive system. Namely, the higher the degree of open innovation of a firm, the higher is the degree of complexity of its complex adaptive system. That the complexity of an innovation system is great is another way of saying that the openness of the open innovation system is great. The complexity and the openness of a specific complex adaptive system are measured by the degree of the open innovation of related firms.

The degree of complexity of a complex adaptive system is shown concretely by the degree of competition between firms or institutions in that system. Emergence at a low level of complexity is the differentiation made by the open innovation of a firm. This differentiation includes not only product differentiation but also price differentiation. If the complexity of a complex adaptive system increases, the dominant design of existing products or the dominant firm itself changes. In addition, if the complexity of the complex adaptive system grows higher, a new innovative sector or industry appears. There are several kinds of complex adaptive systems such as national innovation system, RIS, and SIS. The complexity of the national innovation system as a complex adaptive system varies among countries. The results of corporate open innovation can, in the end, determine the creativeness and complexity of the national innovation system. The differences in the creativeness and complexity among national innovation systems are determined by corporate open innovation, while the degree or level of corporate open innovation is influenced by the institutions that form the national innovation system (Yun et al. 2010, 2012). As a complex adaptive system, the national innovation system is a reflection of the firms, major affiliated agencies, complementarities, self-organization, and proper emergence of the complex adaptive systems in it (Manzini 2009).

Meanwhile, the open innovation of affiliated firms can determine the creativeness and complexity of the belonging RIS. At the same time, the openness and creativeness of the belonging cluster or RIS can determine the degree of open innovation of the affiliated firms (Cooke 2005; Yun and Mohan 2012a; Yun and Choi 2008). For example, in Taiwan, the foundation of the Hsinchu cluster in which knowledge, technology, and capital are free to flow through connections to Silicon Valley is motivated through the activation of diverse open innovation activities of the affiliated firms (Saxenian and Hsu 2005).

Corporate open innovation can increase the creativeness and complexity of a specific sector, while the sectorial innovation system determines the open innovation of the related firms (Yun et al. 2012). Although corporate open innovation affects the improvement of corporate differentiated competitiveness, it increases the complexity and creativeness of the SIS if it is combined with various positive feedback loops, such as economies of scale, network effects, and open innovation business models. If this occurs, the existing dominant design in the related SIS fades away, and fierce competition occurs to establish a new dominant design. Dominant firms of the belonging SIS change rapidly, new markets are set up, or the initial market size and the SIS scope rapidly expand. Ultimately, the sectorial specificities in the geography of a corporate location are determined by corporate open innovation, that is, how the firms in the SIS combine their knowledge, technology, and manufacture of products (Bottazzi et al. 2005).

Complex adaptive systems such as national innovation system, RIS, or SIS lead open innovation of related firms in specific directions, or each has a strange trigger to promote more innovation under specific conditions. The existence of fascinated customers in belonging sectors plays the role of a strange trigger. This induces innovation systems to accept more easily the open innovation made in the sector. Triggering effects may also occur in the case of innovation systems with R&D capabilities and technology. Certain national firms may influence university funding by shaping those institutions, while national university systems crucially affect the competitive advantage of firms in the global market (Francisco et al. 2007). The features and properties of an innovation system act as determinants of acceptance, regarding both the degree and direction of the open innovation of individual firms.

Political intervention by governments is required to enhance the degree of complexity (creativity) or the level of openness in complex adaptive systems or to promote the activation of national innovation system, RIS, and SIS through knowledge production, distribution, and consumption. Government intervention in system failures of complex adaptive systems is aimed at promoting the open innovation of individual firms. Consequently, the core responsibility of government is to build open national innovation system, open RIS, or open SIS, which in turn produces and distributes new knowledge and technology into the innovation system by enhancing the complexity of the complex adaptive system, that is, the openness of the innovation system.

7.3.3 Evolutionary Change in the OCE Model

Evolutionary economics inherited from Schumpeter's legacy involves coevolution of national industries, technology, and institutions such as universities, research labs, and patents (Nelson 1994). The OCE model does not analyze evolutionary results at the level of economy but looks at evolutionary effects reached as a result of differentiation strategies by which open innovation is established at a specific firm. At the current rates of growth in knowledge, rates of its use, and formation of positive feedback loops of new types (e.g., SNS), the selection of open innovation strategies at the corporate level produces rapid evolutionary results in markets. This phenomenon, that is, evolutionary results from open innovation at corporate levels in markets is applicable not only to market-dominating large companies but also to SMEs. Before firms carry forward open innovation strategies, they should check the evolutionary effects of related open innovation strategies that are linked to corporate competitiveness and profits. The time frame from open innovation to evolutionary result is being shortened very sharply. That cutting-edge new technological products face the possibility of falling into the commodity trap in such short times proves the shortened technology life cycle.

An evolutionary model of technological change is proposed in which a technological breakthrough, or discontinuity, initiates an era of intense technical variation and selection, culminating in a single dominant design product (Anderson and Tushman 1990). Namely, the pinnacle of evolution in innovation is the very formation of dominant design. Dominant design goes through a variety of incremental technical progressions according to the differentiation strategies of many firms, by way of open innovation. Dominant design is not fixed. It goes through an evolutionary process created by the open innovation based on discontinuous technology of a firm and then forms another dominant design.

Companies with the best products will not always win, as chance events may cause "lock-in" of inferior technologies (Arthur 1983). The process from dominant design formation to its lock-in is the evolutionary result of open innovation based on new knowledge and technology. Various dynamic evolutionary powers trigger the process from dominant design formation to its lock-in. Like biological evolution, the evolution of whole markets related to technological innovation is never locked-in forever. While switching costs may favor the incumbent during rapid technological change, switching costs can become quickly swamped by switching benefits (Teece et al. 1997). Increasing returns, network effects, economies of scope, and open business models are the forces that enable switching benefits to surpass switching costs.

In economics, positive feedback arises from increasing returns (Arthur 1994). In economies, a positive feedback loop is the driving force making a specific technology win a position of dominant design on the market and then creating lock-in. The market mechanisms that make up this positive feedback loop are economies of scale, economies of scope, economies of network, and an open business model platform.

Economies of scale are a positive feedback loop on the supply side that increases supply so long as profit increases in proportion to the increase in supply. In cases where increasing returns are caused by economies of scale, lock-in for current technology occurs. If one among competing technologies happens to be adopted by historical events, and increasing returns are created through economies of scale, this technology becomes the dominant design and gradually becomes locked-in (Arthur 1989).

Also on the supply side is another positive feedback loop based on economies of scope. It is more efficient for a single supplier to supply a variety of products than for different suppliers to supply those products singly in the same product field. This logic also justifies M&A in microeconomics. When various new types of open innovation occur in a traditional manufacturing industry, they undergo an evolutionary process to dominant design and lock-in through economies of scope. Notably, evolutionary phenomena based on economies of scale or scope loses their power the moment the positive feedback stops. If alternative technology appears suddenly through the open innovation of another firm, and is powered by the positive feedback loop, the existing dominant design can disappear suddenly (Anderson and Tushman 1990).

Economies of network form a positive feedback loop on the demand side. Demand increases geometrically as the bandwagon effect occurs in proportion to the increase in demand. For example, as the number of Microsoft (MS) Office users grows, more users are likely to use the same kinds of documentation. The exchange and distribution of documents then become more convenient, and, accordingly, the number of users increases more. A variety of social network systems (SNS, e.g., Facebook, Twitter, and KakaoTalk) also have positive feedback loops based on typical economies of network. A creative open innovation based on a new idea evolves into a new dominant design if it is powered by a positive feedback loop thanks to economies of network based on fortuitous initial users. This positive feedback loop on the demand side has relatively solid and long-term sustainability. Sales of the QWERTY keyboard have been solidly sustained despite than many more efficient keyboards have been developed since it evolved into the dominant design based on economies of network. Such long sustainability is possible if a new idea reaches the status of dominant design due to economies of network and even a minimum, steady effect from economies of scale is present at some level.

An open business model platform is one with features similar to an evolutionary game. It refers to a phenomenon in which there is a positive feedback loop by which consumers of products turn into producers of product, and then said producers turn back into consumers, now concentrating more on related products. For example, the Apple App Store is an open business model platform in that consumers of apps may turn into producers of apps and then go on to consume more apps. Namely, it is a positive feedback loop in which economies of network coincide with economies of scale on both the demand and supply sides. Apple's iTunes, iBook, and Passbook also have such features of an open business model platform. In the case where firms reach a dominant design through evolution by use of an open business model platform, they have a very solid evolutionary quality even though this may not cause rapid growth like that based on economies of scale (based on supply). Its positive feedback loop is made relatively stable relatively by economies of network (demand), and considerable self-supply occurs simultaneously as some consumers become producers as well. In the positive feedback loop of an open business model platform, consumers turn into suppliers and supply diverse products that are not comparable with those produced due to economies of scope. Consequently, the existence of the long tail phenomenon means that the aggregated quantity of diverse products supplied is greater than that provided by large-scale suppliers. Examples include sales of various T-shirts by Threadless, various book sales by Amazon, and sales of diverse music products by iTunes.

Activation of SNS in forms such as YouTube, Facebook, Twitter, and KakaoTalk activates various new types of positive feedback loops, on both the supply and demand sides. As a typical example, the song "Gangnam Style" by "Psy" collected a massive amount of views through YouTube. Consumers of the song turn into "producers" who put copies of the music video on their own websites. Through the music video, the song rode on a positive feedback loop. In the end, Psy's music video broke the 500 million mark on YouTube in just 91 days, and the song was ranked second on the US Billboard chart. Through activation of SNS, there is rapidly increasing potential for creative open innovation by individuals or firms to be powered by positive feedback loops such that they reach a position of dominant design in a very short time.

7.4 Applying OCE Model

7.4.1 Who Can Use OCE Model?

Most of all, firms can use this model when they build their own innovation strategies. Low level of usage of the OCE model is OCE thinking modeling. If firms can choose an innovation strategy based on the OCE thinking model, they can simulate through mental experiments all the processes including introducing a new innovation strategy, change of the complex adaptive system, and evolutionary change of it. Firms could choose innovation strategies from a dynamic perspective. The middle level of usage of the OCE model is OCE system dynamic modeling. OCE system dynamic modeling means that firms render all the dynamic processes of the OCE model into system dynamics models and apply them to open innovation strategies. High-level usage of the OCE model is OCE agent-based modeling. Firms can make all the OCE processes at agent-based models and apply them to reality when choosing or analyzing strategies (Yun et al. 2016).

In addition to firms, governments or social entrepreneurs can use the OCE model when they choose or build up government policy or social services because government policy or social services are a kind of open connection between technology or knowledge, and society, which is an expansion of the market. From the OCE model perspective, policy agents or social entrepreneurs can forecast or analyze the effects of their policies and social products from a microperspective. The OCE model is thus useful for government and social entrepreneurs together.

7.4.2 How Can We Use the OCE Model?

From concept model building to concrete reality simulation, the OCE model can be used diversely.

First, when any firm builds a new innovation strategy, it can use the OCE model as a concept modeling tool for its new innovation strategy.

Second, consulting groups can make several advanced OCE models that can anticipate the results of choosing different open innovation strategies. For this, several simulation methods such as system dynamics or agent-based modeling can be used.

Third, when researchers or agents evaluate any open innovation strategy or policy after the event, they can use OCE model as an evaluation or analyzing tool. For this, they can use the OCE model from the low level as a thinking experiment tool, through the middle level as a system dynamics model, to the high level as agent-based modeling.

Research Question

1. Consider any open innovation strategy, and apply it the OCE model as a concept model or thinking experiment.
2. Select any open innovation case and apply and analyze it with the OCE Model.
3. Build your own OCE simulation model for your firm or your future start-up, and apply it.

References

Aldrich H (1999) Organizations evolving. Sage, London

Anderson P, Tushman M (1990) Technological discontinuities and dominant designs. Adm Sci Q 35:604–633

Arthur WB (1983) Competing technologies and bock-in by historical events: the dynamics of allocation under increasing returns, Working paper WP. International Institute for Applied Systems Analysis, Laxenburg, pp 83–90

Arthur W (1989) Competing technologies, increasing returns, and lock-in by historical events. Econ J 99(394):116–131

Arthur WB (1994) Increasing returns and path dependence in the ecomomy. University of Michigan Press, Ann Arbor
Arthur WB (1999) Complexity and the economy. Science 284(5411):107–109
Arthur WB (2009) Complexity and the economy. In: Rosser JJ (ed) Handbook of research on complexity. Edward Elgar Publishing, Northampton, pp 12–21
Axelrod R, Dion D (1988) The further evolution of cooperation. Science 242:1385–1390
Bottazzi G, Dosi G, Fagiolo G (2005) On sectoral specificities in the geography of corporate location. In: Breschi S, Malerba F (eds) Clusters, networks, and innovation. Oxford University Press, Oxford, pp 54–79
Capron L, Dussauge P, Mitchell W (1998) Resource redeployment following horizontal acquisitions in Europe and North America, 1988–1992. Strateg Manag J 19(7):631–661
Carcia R, Rummel P, Hauser J (2007) Validating agent-based marketing models through conjoint analysis. J Bus Res 60:848–857
Chesbrough H (2006) Open business model: how to thrive in the new innovation landscape. Harvard Business School Press, Bostom
Chesbrough HW, Teece DJ (2002) Organizing for innovation: when is virtual virtuous? Harvard Business School Publishing, Boston
Coase R (1937) The nature of the firm. Economica 4:386–405
Cooke P (2005) Regional knowledge capabilities and open innovation: regional innovation systems and clusters in the asymmetric knowledge economy. In: Breschi S, Malerba FE (eds) Clusters, networks, and innovation. Oxford University Press, Oxford, pp 80–112
Das T, Teng B (2000) A resource based theory of strategic alliance. J Manag 26(1):31–61
D'Aveni R (2010) Beating the commodity trap: how to maximize your competitive position and increase your pricing power? Harvard Business Press, Boston
Drnevich PL, Kriauciunas AP (2011) Clarifying the conditions and limits of the contributions of ordinary and dynamic capabilities to relative firm performance. Strateg Manag J 32:254–279
Eisenhardt KM, Martin JA (2000) Dynamic capabilities: what are they? Strateg Manag J 21:1105–1121
Fortune A, Mitchell W (2012) Unpacking firm exit at the firm and industry levels: the adaptation and selection of firm capabilities. Strateg Manag J 22:794–819
Francisco F-V, Julio S-C, Jarne G (2007) Modeling the co-evolution of national industries and institutions. Ind Corp Chang 17(1):65–108
Hammond RA, Axelord R (2006) The evolution of ethnocentrism. J Confl Resolut 50(6):1–11
Helfat CE, Finkelstein S, Mitchell W, Peterat MA, Singh H, Teece DJ, Winter WG (2007) Dynamic capabilities: understanding strategic change in organizations. Blackwell, Malden
Heredero C, Berzosa D (2012) Open innovation in firms and public administraions: technologies for value creation. Information Science Reference, Hershey
Jacobides M, Winter S (2005) The co-evolution of capabilities and transaction costs: explaining the institutional structure of production. Strateg Manag J 26:395–413
Kogut B (1988) Joint ventures: theoretical and empirical perspectives. Strateg Manag J 9:319–332
Laursen K, Salter A (2006) Open for innovation: the role of openness in explaining innovation performance among U.K. manufacturing firms. Strateg Manag J 27:131–150
Lee K, Lim C (2001) Technological regimes, catching-up and leapfrogging: findings from the Korea industries. Res Policy 30:459–483
Malerba F, Nelson R, Orsenigo L, Winter S (1999a) History friendly models of industry evolution: the computer industry. Ind Copor Change 1:3–41
Malerba F, Nelson R, Orsenigo L, Winter S (1999b) "History-friendly" models of industry evolution: the computer industry. Ind Corp Chang 8:3–40
Malerba F, Nelson R, Orsenigo L, Winter S (2001) Competition and industrial policies in a 'history friendly' model of the evolution of the computer industry. Int J Ind Organ 19:635–664
Malerba F, Nelson R, Orsenigo L, Winter S (2008) Vertical integration and disintegration of computer firms: a 'history-friendly' model of the co-evolution of the computer and semiconductor industries. Ind Corp Chang 17(2):197–231

Manzini S (2009) Systems, complexity and the national system of innovation: a South African perspective. LAP LAMBERT Academic Publishing, Durban
March J (1991) Exploration and exploitation in organizational learning. Organ Sci 2(1):71–87
Mowery D, Oxley J, Silverman B (1996) Strategic alliances and interfirm knowledge transfer. Strateg Manag J 17:77–91
Nelson R (1994) The co-evolution of technology, industrial structure and supporting institutions. Ind Corp Chang 3(1):47–63
Nelson R, Winter S (1982) An evolutionary theory of economic change. Harvard University Press, Cambridge
Nonaka I (1994) A dynamic theory of organizational knowledge creation. Organ Sci 5:14–37
Porter M (1980) Competitive strategy. Free Press, New York
Porter M (1985) Competitive advantage. Free Press, New York
Porter M (1990) The competitive advantage of nations. Free Press, New York
Potts J (2001) Knowledge and market. J Evol Econ 11:413–431
Press WH, Dyson FJ (2012) Iterated prisoner's dilemma contains strategies that dominate any evolutionary opponent. PNAS (Proc Natl Acad Sci USA) 09(26):10409–10413
Room G (2011) Complexity, institutions and public policy. Edward Elgar Publishing, Northampton
Saxenian A, Hsu J-Y (2005) The Silicon Valley-Hsinchu connection: technical communities and industrial upgrading. In: Breschi S, Malerba FE (eds) Clusters, networks, and innovation. Oxford University Press, Oxford, pp 235–260
Schumpeter J (1934) Theory of economic development. Harvard Universityh Press, Cambridge, MA
Schumpeter J (1942) Capitalism, socialism, and democracy. Harper, New York
Shuen A (1994) Technology sourcing and learning strategies in the semiconductor industry. Unpublished PhD dissertation, University of California, Berkeley
Simon H (1991) Bounded rationality and organisational learning. Organ Sci 2:125–134
Simon HA (1995) Artificial intelligence: an empirical science. Artif Intell 77:95–127
Teece D, Pisano G (1994) The dynamic capabilities of firms: an introduction. Ind Cop Change 3(3):537–556
Teece D, Rumelt R, Dosi G, Winter S (1994) Understanding corporate coherence: theory and evidence. J Econ Behav Organ 23:1–30
Teece D, Pisano G, Shuen A (1997) Dynamic capabilities and strategic management. Strateg Manag J 18(7):509–533
Tunzelmann NV, Wang Q (2007) Capabilities and production theory. Struct Chang Econ Dyn 18:192–211
Wilensky U (2006) Netlogo ethnocentrism model. http://ccl.northwestern.edu/netlogo/modles/Ethnocentrism, Center for Connected Learning and Computer-Based Modeling
Williamson O (1991) Strategizing, economizing, and economic organization. Strateg Manag J 12:74–95
Won D, Yoo S, Yoo H, Lim J (2015) Complex adaptive systems approach to sewol ferry disaster in Korea. J Open Innov Technol Market Complex 1(1):1–18
Yoon M, Lee K (2009) Agent-based and 'history-friendly' models for explaining industrial evolution. Evol Inst Econ Rev 6(1):45–70
Yun JJ (2015) How do we conquer the growth limits of capitalism? Schumpeterian dynamics of open innovation. J Open Innov Technol Market Complex 1(1):1–20
Yun JJ, Choi M (2008) A study on the difference of the relationship between open innovation and firm performance according to the clusters. Korea Policy Assoc J 17(4):163–191
Yun JJ, Mohan AV (2012a) The relation between internal and external open innovation: a study of firms located in the Goomi and Banwol-sihwa clusters in South Korea. In: Brem A, Tidd JE (eds) Perspectives on supplier innovation: theories, concepts and empirical insights on open innovation and the integration of suppliers. Imperial College Press, London

Yun J, Mohan A (2012b) Exploring open innovation approaches adopted by small and medium firms in emerging/growth industries: case studies from the Daegu-Gyeongbuk region of South Korea. Int J Technol Policy Manage 12:1–19
Yun JJ, Park S, Lim D, Hahm S (2010) Emergence of East Asian TFT-LCD clusters: a comparative analysis of the Samsung cluster in South Korea and the Chimei cluster in Taiwan. Asian J Technol Innov 18(1):201–228
Yun JJ, Nadhiroh IM, Jung WY (2012a) A study on the relationship between open innovation, Entrepreneurship and business model from Korean and Indonesian IT SMEs. Asialics 9th: innovation and appropriate technology for the development and inclusive growth of Asian SMEs, Manila: Asian Network for Learning, Innovation, and Competence Building systems, 7
Yun JJ, Ryu GW, Jung WY (2012b) Is there any difference in the effect of open innovation according to the product life cycle?: thinking again product life cycle and trade. Korea Trade Research Association 2012 October conference. Korea Trade Research Association, SokCheo, pp 321–350
Yun JHJ, Won DG, Park KB (2016) Dynamics from open innovation to evolutionary change. J Open Innov Technol Market Complex 2(7):1–22
Zollo M, Winter SG (2002) Deliberate learning and the evolution of dynamic capabilities. Organ Sci 13(3):339–351

Part IV

Relation Between Open Innovation and Business Model

The Relationship Between Open Innovation, Entrepreneurship, and Business Model

8

Abstract

This chapter looks into the relationship among open innovation (OI), entrepreneurship, and introduction of a new business model (BM) based on case studies of information technology firms in Daegu and Seoul in Korea and in Indonesia. From this, we develop a perspective about the relationship between open innovation and in two different contexts: between different regional innovation systems (RISs) and different national innovation systems (NISs).

Keywords

Open innovation • Entrepreneurship • New business model • National innovation system • Regional innovation system

8.1 Introduction

Open innovation is essential for SMEs that cannot afford their own research and development or that lack the expertise to develop their own products (Yun and Mohan 2012). Especially in sectors where the market changes rapidly and technology advances quickly, SMEs depend on various forms of open innovation, such as user innovation or demand heterogeneity, rather than innovation based on their own

---

This chapter is mainly based on the following paper. More than 30% of the paper was changed at this paper. Please read the full paper for the understanding of this chapter totally.
Yun J.H.J, Nadhiroh I.M, Jung W.Y (2013), The relationship between open innovation, entrepreneurship, and introduction of new business models in Korean and Indonesian information technology enterprises. *Korean Social Science Journal*. Vol. 40, No. 2, pp 81–99.

J.J. Yun, *Business Model Design Compass*, Management for Professionals,
DOI 10.1007/978-981-10-4128-0_8

research and development (Yong and Park 2010). Smaller enterprises are taking an increasingly prominent role in the contemporary innovation landscape (Varande et al. 2008).

This chapter presents case studies on information technology (IT) SMEs to determine whether there is a concrete relationship between open innovation, entrepreneurship, and business model at the enterprise level. In addition, we analyzed the difference in this relation between RISs and national innovation systems to find additional implications.

Normally, the research methods used for studying the open innovation of SMEs are statistical analysis based on surveys and individual case analysis. Statistical analysis based on surveys divides open innovation channels of SMEs into two categories: (1) technology exploitation, based on venture building outward industrial property licensing and employee involvement and (2) technology exploration, based on customer involvement, external networking, external participation, outsourcing research and development (R&D), and inward industrial property licensing. It then analyzes surveys to identify the difference between manufacturing and service industries (Varande et al. 2008). In addition, many studies, such as a recent study on SME innovation strategy (Lecocq and Demil 2006), a study on informal collaboration in open source software development (Henkel 2006), and a study on SME strategies for searching for external knowledge (Laursen and Salter 2006), analyze the open innovation of SMEs in depth and breadth using a survey method. The survey method is useful for establishing a theory or generalizing a phenomenon, but has limited usefulness for identifying the characteristics of open innovation by individual SMEs.

Many other studies used case studies where certain characteristics of SMEs are deduced by interviewing many SME staff members, implicitly or explicitly, using a semi-structured questionnaire or checklist for interviews (Massa and Testa 2008). Another case study analyzed the role of technology in the process of changing the basic policy of individual enterprises to an open innovation strategy, although that was for just a single enterprise, Procter & Gamble. The case study was conducted based on an interview template, with members of many departments as the subjects (Dodgson et al. 2006).

We used the case study method. We made a checklist (shown in the Appendix) for an interview to investigate the specific circumstances of open innovation, entrepreneurship, and new business model introduction of IT SMEs. In addition, we employed the analytic hierarchy process (AHP) method to assess the degrees of the three factors at each enterprise.

All interviews were conducted between May and September 2011. The Korean research team interviewed the Korean enterprises, and the Indonesian team interviewed the Indonesian enterprises. With a single semi-structured questionnaire, a consensus on significance, interpretation, and summary of interview results was formed among the research teams. The study results were compared and analyzed through a series of discussions after the interviews.

8.2 The Relation Among Three Factors

8.2.1 The Relationship Between Open Innovation and Introduction of a New Business Model

Innovative performance resulting from an enterprise's open innovation can be categorized into three types: products new to the world market, products new to the enterprise, and significantly improved products for the own firms (Laursen and Salter 2006). Open innovation channels—such as cooperation among Chinese SMEs, cooperation with intermediary institutions, and cooperation with research organizations—have been shown to have positive effects on innovation performance, such as the annual turnover of new products or the innovation index (Zeng et al. 2009). Similarly, collaboration with partners in the value chain (customers or suppliers) provides a strong base for the incremental improvement of existing products and services, whereas collaboration with academic institutions increases the ability of enterprises to drive radical new product development, due to access to new technologies (Parida et al. 2012). With the exception of analyses of individual open innovation cases, most open innovation studies, like the cases above, analyze how much the results of open innovation affected the introduction of new products, which is typically reflected as sales or the proportion of sales.

The introduction of a new business model indicates new products and new services such as A in Fig. 8.1. In other words, the measures of performance of open innovation through the interviews were the presence, degree, and frequency of new business model introduction and the qualitative aspects of new product introduction.

8.2.2 The Relationship Between Entrepreneurship and Open Innovation

The entrepreneurship fuzzy set has three dimensions in the market sector: organization creation, economic innovation, and profit seeking (Hornaday 1992; Kim and Jung 2015). Among these, organization creation and economic innovation are associated with the organizational behavioral characteristics of enterprises related to open innovation. SME entrepreneurs have strong viewpoints about innovation, including that "innovation is anything that makes money; innovation comes from everyone; and innovation sources are everywhere" (Massa and Testa 2008, p. 409). SMEs tend to consider strict rules to be obstacles to innovation and focus on the importance of research centers and universities and intermediaries that mediate technology and ideas. Schumpeterian entrepreneurs monitor developments in technologies, products, and methods at home and abroad and contemplate how profitable it would be to adapt or improve existing goods or methods or produce them less expensively (Phelps and Zoega 2009).

Conceiving of new products and new methods against the background of existing technologies and the accessible stock of past products and methods is generally the

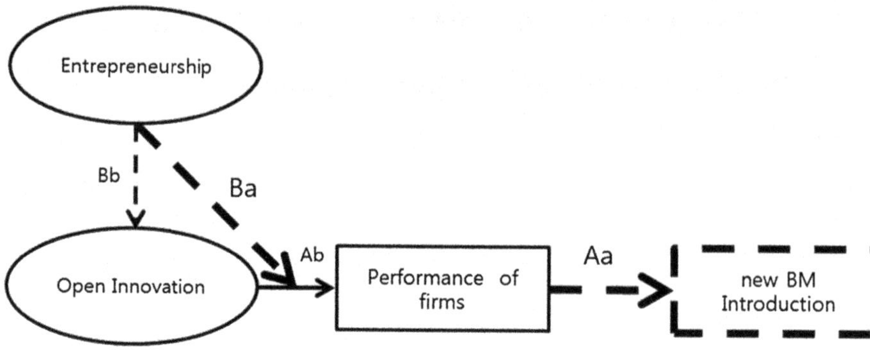

Fig. 8.1 The relation between entrepreneurship, open innovation, and business model (Source: Yun et al. (2013) modified)

contribution of entrepreneurs (Hayek 1978). In addition, this entrepreneurial orientation has moderating effects on the market orientation–performance linkage, according to evidence from Chinese small enterprises (Li et al. 2008). Open and networked innovation that is triggered by a creator with visionary leadership, such as entrepreneurship, is a process of exploration and exploitation of knowledge (Harryson 2008).

Entrepreneurship plays an important role in promoting open innovation. In fact, the distinctive appearance of entrepreneurship stimulates the system, equipment, and enterprise culture, laying the groundwork for the introduction of a larger degree of open innovation, such as Ba and Bb in Fig. 8.1.

8.2.3 The Relation Between Open Innovation and RIS or National Innovation System

An appealing attribute of clusters or RISs is that they can provide positive externalities to agglomerated enterprises because they are interconnected and would theoretically encourage information and collaboration flows among members (Silvestre and Dalcol 2009; Marceau 1994). Within a cluster, the intensity of information and communication is observed through face-to-face contacts and the interactions of people and enterprises (Batheld et al. 2004).

However, because there is a severe asymmetry among clusters, depending upon their makeup and location, some locations or clusters in any industry or sector have more knowledge than others (Malmberg 2003). For example, the difference in production, distribution, and consumption among growing clusters may lead to a difference among specific enterprises, such as Samsung Display Ltd. and Chimei Display Ltd. among Korean and Taiwan thin-film-transistor liquid-crystal display (TFT LCD) clusters (Yun et al. 2010). Differences between RISs and clusters cause an unequal distribution of knowledge within the economic system and lead to a difference in regional knowledge capabilities and open innovation between an RIS and a cluster (Cooke 2005). A study that analyzed the Silicon Valley–Hsinchu Connection demonstrated that as a new cluster became affiliated with the Silicon Valley through the construction of a Taiwanese technical community there, the new cluster became successful due to an increase in the mobility of knowledge and capital (Saxenian and Hsu 2005). Another study found that, according to its regional innovation or regional embeddedness, the mode of knowledge sourcing of an enterprise can be changed (Kramer and Diez 2011).

In addition, there is a difference in mobility and existence of knowledge among regions beyond the boundaries of countries. Meanwhile, differences between national innovation systems, which would include "all important economic, social, political, organizational, institutional, and other factors that influence the development, diffusion, and use of innovation," combined with differences among sectors, lead to a difference in the amount of knowledge, as well as the distribution or circulation speed of knowledge and information, and the commercialization pattern of knowledge between countries surrounding a specific sector (Lundvall 1992, p. 78). The reason is that differences between national innovation systems reflect differences in innovative capacity (Freeman 1987). Another study suggested that, according to different nations' RISs, there will be differences in the intangible assets that act as drivers of innovation (Kramer et al. 2011). Those intangible assets are the amount, speed, or content of knowledge.

Differences between RISs, clusters, and national innovation systems cause differences in the amount of knowledge, as well as its distribution speed, within the relevant innovation system. They consequently induce overall differences among firms in belonging innovation system in terms of the open innovation effect (A), the direct effect of entrepreneurship to open innovation activity (Ba), and the indirect effect to open innovation results through moderating of open innovation (Bb).

8.2.4 The Relation Among Three Factors

Entrepreneurship normally has an indirect effect on the open innovation strategy of firms, such as Ba in Fig. 8.1. Through this, entrepreneurs let their firms increase or activate open innovation and thereby increase performance. But entrepreneurship and open innovation are different in terms of identity. Entrepreneurship is characteristics of the builder of big enterprises, SMEs, start-ups. In contrast open innovation is a kind of strategy employed by firms.

Open innovation motivates increased performance in a target firm. In the end, the increased performance of any firm that is motivated by open innovation introduces a new business model, such as Aa in Fig. 8.1.

The concrete relation among entrepreneurship, open innovation, and business model should be analyzed. It is possible that the concrete relations will be different among RIS, sectorial innovation system (SIS), and national innovation system.

8.3 Difference of Three Factors According to RISs and National Innovation Systems

8.3.1 The Difference Between Seoul RIS and Daegu RIS in Korea

Entrepreneurs in two enterprises in the Seoul RIS (enterprises A and D) and two enterprises in the Daegu RIS (enterprises G and J) established systems for open innovation, such as the three-channel system or recruiting people from a target area. In other enterprises in the Seoul RIS and one enterprise in Daegu, open innovation was actively triggered by entrepreneurs. Some entrepreneurs organized an open innovation system. Others triggered open innovation activity, while others did not encourage open innovation. The open innovation of the Seoul RIS is higher than that of the Daegu RIS, as seen in Fig. 8.2.

From this, the researchers found that the open innovation strategy of an enterprise is directly dependent on entrepreneurship. If the entrepreneurship level of an enterprise is not high—for example, if the CEO does not have a positive attitude toward change, risk, or introducing a new business model—then open innovation in the enterprise is passive and no open innovation-related system will be organized, as was the case with enterprise K. But if the entrepreneurship level of an enterprise is high—if the CEO has a positive attitude toward change, risk, or introducing a new business model—then the open innovation of the enterprise is positive, and an open innovation system often will be organized in the enterprise, as occurred in a particularly energetic way in enterprise A.

As shown in Fig. 8.2, enterprises A, B, and E in the Seoul RIS and enterprise G in the Daegu RIS have open innovation channels to a university or national laboratory, through a supply enterprise, or from the customer sector they belong to. These enterprises have diverse open innovation channels and receive qualitatively different knowledge from these channels. Enterprises C and D in the Seoul RIS and enterprises F, J, and K in the Daegu RIS have open innovation channels mainly to universities or national laboratories, supply enterprises, or their own sector. These enterprises received a lot of knowledge from these channels but did not have serious differences between them. More than half of the Seoul RIS enterprises have active open innovation channels to customers.

From this, we found three levels of open innovation: low (which mainly has a university or national laboratory as the open innovation channel, such as enterprise H and I), middle (which has a university in addition to the enterprises' own sector or competing enterprises as open innovation channels, such as enterprises C, D, K, J, and K), and high (which has a university, the enterprise's own sector, and customers as open innovation channels, such as enterprises A, B, E, and G).

As shown in Fig. 8.2, enterprises A and D in the Seoul RIS and enterprises G and H in the Daegu RIS have expanded existing business models and at least partially evolved to a qualitatively different business model. Enterprises B, C, and E in the Seoul RIS and enterprise H in the Daegu RIS expanded their business models widely. More than half of the Seoul enterprises expanded their business models in terms of both quantity and quality.

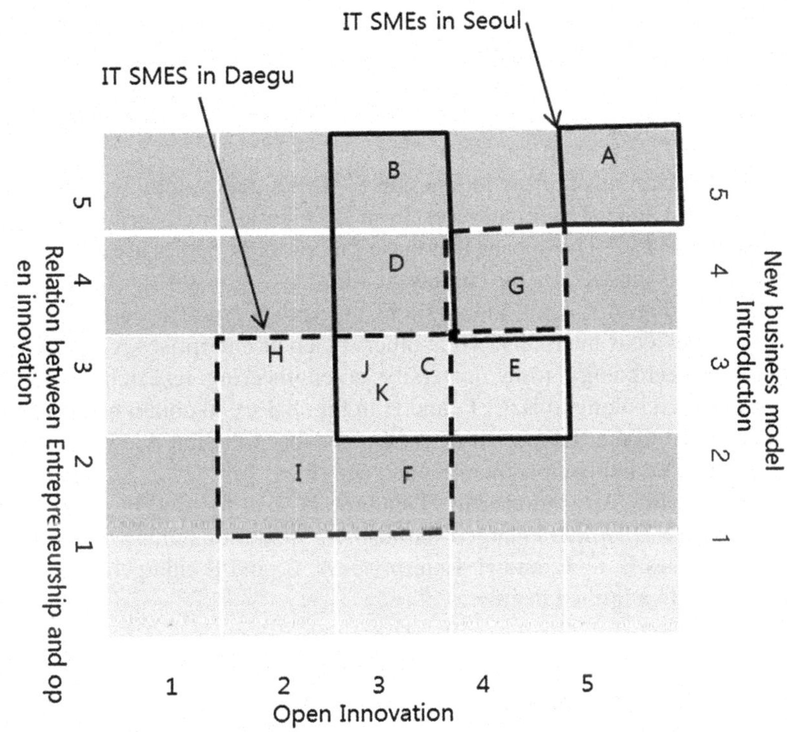

Fig. 8.2 The differences between Daegu RIS and Seoul RIS (Source: Yun et al. (2013))

From this, we identified three levels of new business model introduction: low (adhering to an existing business model without trying to change or enlarge it, such as enterprises F, I, and K), middle (enlarging a business model quantitatively, such as enterprises B, C, E, and H), and high (enlarging a business model quantitatively and evolving it qualitatively, such as enterprises A, D, G, and J).

The differences in these three aspects between RISs were identified based on five cases in one RIS and six cases in the other. The realistic qualitative differences between RIS can be seen through this case analysis. If the differences between level of open innovation and level of relationship between entrepreneurship and open innovation can be addressed, then the introduction of new creative business models will become different according to the RISs to which they belong.

8.3.2 The Difference Between Korean National Innovation System and Indonesia National Innovation System

Indonesia's five IT enterprise CEOs are engineers. In this case, when CEOs try to develop new ideas by themselves, such as in enterprise B, the relationship between entrepreneurship and open innovation is very low. But if CEOs try to let their engineers develop new ideas, such as in enterprises A, C, and E, the relationship between entrepreneurship and open innovation is higher. Enterprise C tries to invite and implement unusual ideas from outside the enterprise or from the gaming community.

The new ideas in most of the Indonesian IT SMEs under study were introduced by an engineer-CEO or by employees from an internal engineering department, except in enterprise C. This means that these Indonesia IT SMEs use a closed innovation strategy as directed by the engineer-CEO.

As shown in Fig. 8.3, some Indonesia IT enterprises (such as enterprises B and C) generate new ideas internally, while others (such as enterprises A and D) obtain new ideas or technology from university or engineering research laboratories. Enterprises which belong to B, C, E, and H in Fig. 8.3 try to obtain new ideas open innovation. Most of the Indonesian IT enterprises that were interviewed have internal or limited open innovation channels and activities.

As shown in Fig. 8.3, Indonesian IT enterprises had focused business models, except enterprise D, which enlarged its business model quantitatively, much like Korean enterprises B, C, E, and H. Enterprises A, C, and E enlarged their original business model to a limited degree.

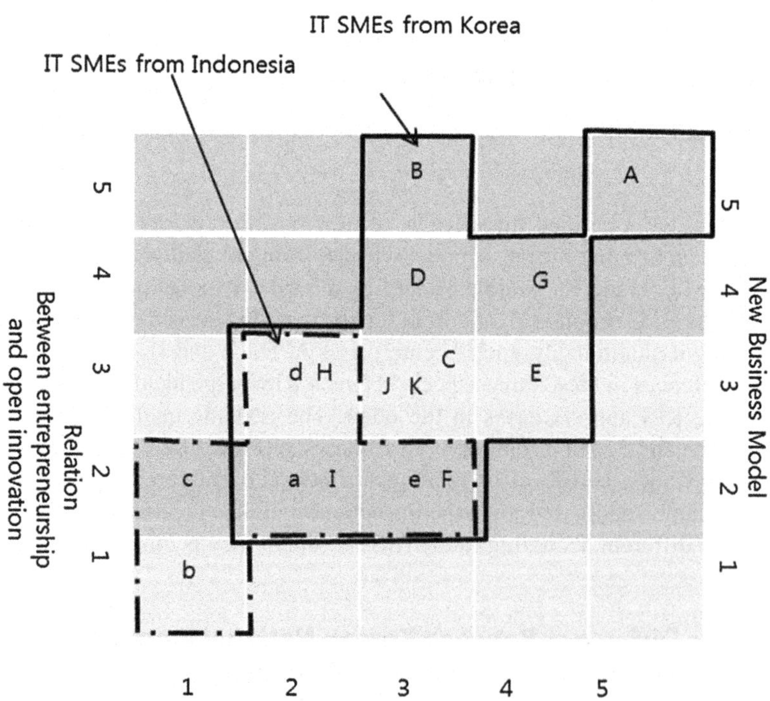

Fig. 8.3 The differences between Korea National Innovation System and Indonesia National Innovation System (Source: Yun et al. (2013))

From this, we could conclude that any enterprise with closed innovation cannot enlarge its business model quantitatively or evolve its business model qualitatively. But this does not exclude the possibility of closed innovation by the enterprise itself.

When comparing Indonesian IT SMEs to those of Korea, it is most of all apparent that the levels of relationship between entrepreneurship and open innovation, open innovation, and introduction of new business models are all lower in Indonesia. The level of IT SMEs in Indonesia is low overall—except for some, like enterprise E, which has open innovation activities for the acquisition of new ideas and maintains internal and external systems. Except for enterprise D, entrepreneurs themselves showed little openness toward external ideas and little direct interest in building and maintaining an open system. Even considering that the IT enterprises of Indonesia are in their initial stages, the new business model aspect of these enterprises is insufficiently activated, with a few exceptions. It is concluded that the new business model aspect remains at the stage of a supply-oriented limited product business model in Indonesia's IT sector (Fig. 8.3).

However, the IT SMEs of Indonesia are basically promoting a closed innovation strategy centered on technical innovations. They are expanding technology through research collaboration with universities or research centers and other outside enterprises within a limited range. Because the IT sector has not yet fully developed or matured, it is concluded that the enterprises in Indonesia have not reached the stage of creative open innovation that is based on demand, expectations, and ideas from customers.

All five enterprises in Indonesia were founded by engineers—unlike in Korea, where many IT SMEs do not have engineers as founders or CEOs. Even if this was not the case, by securing senior executives from the management or marketing sectors, the Korean enterprises tend to promote a market-oriented open innovation strategy and management. Engineer-oriented Indonesian enterprises are promoting technology-push-based business management, which is also internally based on engineers. Therefore, their diverse market and external open innovation idea procurement and new business model introduction still seem somewhat insufficient.

Unlike in Indonesian IT SMEs, the level of open innovation in Korean IT SMEs is fairly high. In that situation, various levels of open innovation have become a source for procuring new business models. Above all, entrepreneurship that is open to new markets is considered to be very important to the expansion of open innovation. Therefore, Korean IT SMEs at their current stage are considered to be in a situation where the frequent introduction of new business strategies determines an enterprise's competitiveness by various channels, and ways of open innovation with more open entrepreneurship serve as their base.

Research Question

1. Select any firm, and look into the relation between open innovation and entrepreneurship.
2. Select any firm, and look into the relation between open innovation and business model.

References

Batheld H, Malmberg A, Maskell P (2004) Cluster and knowledge: local buzz, global pipelines and the process of knowledge creation. Prog Hum Geogr 28:31–56

Cooke R (2005) Regional knowledge capabilities and open innovation: regional innovation systems and clusters in the asymmetric knowledge economy. In: Breschi S, Malerba F (eds) Clusters, networks, and innovation. Oxford University Press, Oxford, pp 80–112

Dodgson M, Gann D, Salter A (2006) The role of technology in the shift towards open innovation: the case of Procter & Gamble. R&D Manag 36(3):27–35

Freeman C (1987) Technology policy and economic performance: lessons from Japan. Pinter, London

Harryson S (2008) Entrepreneurship through relationships—navigating from creativity to commercialisation. R&D Manag 38(3):290–310

Hayek FA (1978) Competition as a discovery procedure. In: Hayek FA (ed) New studies in philosophy, politics, economics and the history of ideas. Routledge, London, pp 179–190

Henkel J (2006) Selective revealing in open innovation processes: the case of embedded Linux. Res Policy 35(7):935–969

Hornaday RW (1992) Thinking about entrepreneurship: a fuzzy set approach. J Small Bus Manag 30(4):12–23

Kim J-h, Jung S-h (2015) Study on CEO characteristics for management of public art performance centers. J Open Innov Technol Market Complex 1(1):1–21

Kramer J-P, Diez J (2011) Catching the local buzz by embedding? Empirical insights on the regional embeddedness of multinational enterprises in Germany and the UK. Reg Stud 15(2):1–15

Kramer J-P, Marinelli E, Iammarino S, Diez J (2011) Intangible assets as drivers of innovation: empirical evidence on multinational enterprises in German and UK regional systems of innovation. Technovation 31:447–458

Laursen K, Salter A (2006) Open for innovation: the role of openness in explaining innovation performance among U.K. manufacturing firms. Strateg Manag J 27(2):131–150

Lecocq X, Demil B (2006) Strategizing industry structure: the case of open systems in a low-tech industry. Strateg Manag J 27(9):891–898

Li Y, Zhao Y, Tan J, Liu Y (2008) Moderating effects of entrepreneurial orientation on market orientation-performance linkage: evidence from Chinese small firms. J Small Bus Manag 46(1):113–133

Lundvall B-A (1992) National system of innovation: towards a theory of innovation and interactive learning. Pinter, London

Malmberg A (2003) Beyond the cluster-local milieus and global connections. In: Peck J, Weung W (eds) Remaking the global economy: economic-geographical perspectives. Sage, Thousand Oaks, pp 256–275

Marceau J (1994) Clusters, chains and complexities: three approaches to innovation with a public policy perspective. In: Dogson M, Rothwell R (eds) The handbook of industrial innovation. Edward Elgar, Cheltenhan, pp 145–171

Massa S, Testa S (2008) Innovation and SMEs: misaligned perspectives and goals among entrepreneurs, academics, and policy makers. Technovation 28:393–407

Parida V, Westerberg M, Frishammar J (2012) Inbound open innovation activities in high-tech SMEs: the impact on innovation performance. J Small Bus Manag 50(2):283–309

Phelps E, Zoega G (2009) Entrepreneurship, culture and openness. In: Audretsch DB, Litan R, Strom R (eds) Entrepreneurship and openness: theory and evidence. Edward Elgar, Northampton, pp 101–121

Saxenian A, Hsu J-Y (2005) The Silicon Valley-Hsinchu connection: technical communities and industrial upgrading. In: Breschi S, Malerba F (eds) Clusters, networks, and innovation. Oxford University Press, Oxford, pp 230–255

Silvestre BD, Dalcol PR (2009) Geographical proximity and innovation: evidences from the Campos Basin oil & gas industrial agglomeration. Technovation 29:546–561
Varande VV, Jong JP, Vanhaverbeke W, Rochemont MD (2008) Open innovation in SMEs: trends, motives and management challenges. Technovation 29:423–437
Yong SS, Park KD (2010) The role of demand heterogeneity in product innovation strategy. Korean Soc Sci J 37(2):27–42
Yun JJ, Mohan AV (2012) Exploring open innovation approaches adopted by small and medium firms in emerging/growth industries: case studies from Daegu-Gyeongbuk region of South Korea. Int J Technol Policy Manage 12(1):1–19
Yun JJ, Park S, Lim D, Hahm S (2010) Emergence of east Asian TFT-LCD clusters: a comparative analysis of the Samsung cluster in South Korea and the Chimei cluster in Taiwan. Asian J Technol Innov 18(1):201–228
Yun JHJ, Nadhiroh IM, Jung WY (2013) The relationship between open innovation, entrepreneurship, and introduction of new business models in Korean and Indonesian information technology enterprises. Korean Soc Sci J 40(2):81–99
Zeng S, Xie X, Tam C (2009) Relationship between cooperation networks and innovation performance of SMEs. Technovation 30:181–194

Business Model and Open Innovation Conditions for the Sustainable Growth of SMEs

9

Abstract

The present study is intended to define and clarify the necessary open innovation and business model conditions for the sustainable growth of SMEs (small and medium-sized enterprises). Initially, we create the open innovation and business model conditions for the sustainable growth of SMES. We devise four different conditions for effective business model and knowledge strategies. The highest growth limits of the four conditions were estimated and then measured for 27 SMEs in the area of IT (information technology) in Korea. From this, we attempt to understand the relationship between open innovation and business model in the context of the growth limits of SMEs. Using this context, we can conceive of the relationship between open innovation and business model in a broader and more concrete manner.

Keywords

Business model • Knowledge strategy • Open innovation

This chapter is based on the following paper. But, this chapter includes just nearly 65% of the paper. Please read the following full paper.
Yun, J.H.J, Jung W.Y, Yang J.H. (2015), 'Knowledge strategy and business model conditions for sustainable growth of SMEs', Journal of Science & Technology Policy Management, Vol. 6, No. 3, pp. 246–262.

J.J. Yun, *Business Model Design Compass*, Management for Professionals,
DOI 10.1007/978-981-10-4128-0_9

9.1 Introduction

Nelson and Winter (1982) developed an evolutionary theory of the capabilities and behaviors of business firms operating in a market environment and constructed and analyzed a number of models consistent with their theory.

9.1.1 Why Do Some Firms Grow Sustainably and Others Disappear in the Short Term?

Among SMEs, some firms develop new products and grow dramatically, whereas other firms, despite the fact that they may have once received attention from markets and competing firms owing to their innovative and customer-attracting products, lost their market power and disappeared in 3–4 years. Why do some firms continuously develop while other firms collapse so soon in this highly competitive environment, with shortened technology life cycles and a widely spreading commodity trap (Aveni 2010)?

In particular, in the IT industry, where the speed of technological advancement is high and product life cycles are very short, clear contrasting cases can easily be found among SMEs, such as firms that grow and develop and those that shrink and become extinct in a short period of only 3–4 years. The factors that determine the growth and development of firms are quite diverse. On the other hand, technology that pushes based on the supply of technology does not always lead to the growth and development firms' (Peters et al. 2012).

First, we will analyze existing studies of the knowledge strategies and business models of firms and the relationship between the two (Morris et al. 2005; Wenger 2004).

Second, we will create the framework between business models and open innovation as one type of reference as we attempt to understand the relationship between open innovation and business models.

Third, 27 firms are concretely analyzed by interviews based on the framework. The present state of sustainable growth of the 27 firms and the conditions and elements that determine the growth are elucidated through this framework.

In this chapter, factors for the sustainable growth of firms were identified based on interviews with the owners of firms (Creswell 2012; Seidman 2006). This chapter includes narrative and phenomenological research which is essentially based on the case study research method. We investigated contemporary phenomena within a real-life context because the boundaries between the phenomena and the context are not clearly evident, and multiple sources of evidence are needed (Yin 2008). The research was carried out via semi-structured, in-depth interviews based on a half-structured questionnaire including questions about the knowledge strategy of firms, the number of years of establishment, the present states and changes of the business models, changes of employees, revenue amounts, structures, and trends.

However, firms' performances and records were obtained by accessing published financial statement databases and through concrete data on firms' strategies, or by complementarily examining business models through homepage searches, press report content examinations, and similar means.

9.2 The Relationship Between Business Model and Open Innovation

9.2.1 In Advance Discussions, Knowledge Strategies, and Business Models

First, we review knowledge strategies. The knowledge management literature highlights the fact that, in the new economy, the achievement of a sustained competitive advantage depends on a firm's capacity to develop and deploy its knowledge-based resources (Carrillo 2015; Ordóñez de Pablos 2002). Knowledge management and organizational learning are clearly positively related (Ordóñez de Pablos 2002). Evidence of intellectual capital measurements from Asia, Europe, and the Middle East provide concrete evidence of this (Ordóñez de Pablos 2002).

Firms acquire external knowledge to avoid having to reinvent the wheel (Lopez and Esteves 2012). The relationship between external and internal networks acts as an enabler of knowledge acquisition and appropriation. Connections to technologi cal partners as providers of knowledge through external networks and the acquisition of knowledge and expertise on how to manage it through internal networks play core roles in the survival of firms. Externally, firms' competitive advantages depend on their ability to use interfirm collaboration to access the essential knowledge and specialized capabilities obtained by other companies, which are difficult to imitate or acquire (McEvily and Marcus 2005). Therefore, firms' knowledge strategies have been recognized as a core element of their sustained survival and development by many researchers (Carrillo and Batra 2012).

Firms as knowing organizations postulate learning and innovation by managing holistically the activities of sensemaking, knowledge creation, and decision-making (Choo and Johnston 2004; Yun and Ryu 2012). In the organizational knowledge cycles, firms estimate demand levels in business environments through sensemaking and discern and fill gaps between external and internal knowledge to create further knowledge. Discussion about knowing organizations also focus on their continuous acquisition of knowledge for sustained growth and development.

The self-efficacy theory provides a unique theoretical model that illustrates how individuals may be motivated to share complex, tacit knowledge in an open source community (Endres et al. 2007). By comparing individuals' self-efficacy to share complex tacit knowledge in the open source versus traditional organizational contexts, researchers can add to the knowledge of how and why knowledge sharing occurs. The self-efficacy theory elucidates the basic elements of firms' sustained development through knowledge strategies by defining motives at the individual level and connection relationships for the acquisition and diffusion of tacit knowledge within firms.

Methods to enhance the transfer of knowledge from baby boomers to Generation X aerospace engineers require visible and participative management involvement. In such a situation, management support is the core of a knowledge-sharing culture that fosters open and honest communication, respectful and trusting relationships, effective mentoring relationships, dynamic team environments, colocations of team members, and a technology infrastructure (McNichols 2010). That is, for sustained development, firms require a team effort for knowledge transfer between generations and across time. The knowledge transfer model, depicting optimal knowledge transfer methods and strategies as identified by Generation X aerospace engineers, is a knowledge-sharing culture.

Firms must begin to create worker-centered environments to encourage the open sharing and use of all forms of knowledge, such as tacit knowledge based on common sense and explicit knowledge based on academic prowess (Smith 2001; Yun and Mohan 2012a, b). Knowledge-creating companies that transform data or tacit knowledge into valuable knowledge can develop sustainably. Organizations that recognize and use their employees' steadily growing tacit and explicit knowledge to solve problems and achieve goals bring a major competitive advantage (Smith 2001).

A macro-process of knowledge management for continuous innovation is required (Xu et al. 2010). The internalization phase of continuous innovation through knowledge management of artifacts, human resources, and technologies are required at the firm level. For continuous innovation and development of firms, a macro-process of knowledge management is required for new knowledge during the innovation process occurring at the contact point between knowledge creation and knowledge usage.

From the foregoing discussion, the importance of knowledge strategies, including those pertaining to the acquisition, management, and development of knowledge as the driving force of sustainable growth and development not only at the individual level but also at the firm level, could be identified. That is, the effective and continuous acquisition and the creation of knowledge form the core strategies of sustained growth by firms.

Second, the business model deserves a discussion. No generally accepted definition of the concept of "business model" has been devised. A six-component framework is proposed for characterizing a business model regardless of the venture type, including factors related to the offering, market, internal capability, competitive strategy, economic, and personal or investor factors (Morris et al. 2005). A business model as "an architecture of the product, service and information flows including a description of the various business actors and their roles, a description of potential benefits for various business actors, a description of the sources of revenues" has the same meaning and context (Timmers 1998). This concept of a business model is clearly distinguished from technology, knowledge, or knowledge strategies. Of course, in terms of value creation for consumers, it is also distinguished from technological innovations by society as addressed in the sociology of science (Yun et al. 2011).

According to an e-business model ontology for modeling e-business, an e-business model framework has three factors, the customer relationship, product innovation, and infrastructure management (Osterwalder and Pigneur 2002). Business models in e-business are eventually shown to have characteristics distinguished from those of knowledge strategies because they do not refer to knowledge or technology per se but instead focus on consumer relations or at least products. That is, according to this line of thought, business models should be viewed separately from knowledge as a factor for the sustained growth of firms.

According to another theory, the elements of a successful business model include customer value propositions (CVPs), profit formulas, key resources, and key processes, and among these, CVPs is the key factor (Johnson et al. 2008). The core reason for the success of Apple despite the fact that it released digital music players to the market later than Diamond Multimedia released in 1999 or Cabo 64 in 2000 is that CVPs were epochally improved by making the downloading of digital music easy. In this discussion, technology or knowledge is included in business models as one of the key resources, but only to a limited extent because CVPs that are finally implemented through key resources, key process, and profit formulae are regarded as business models. Therefore, business models as discussed here should also be considered and reviewed separately from technology or knowledge as a factor for the sustainable growth and development of firms.

Business model innovations are no longer only about technology (Chesbrough 2007). According to Chesbrough (2007), the functions of a business model have six points (value proposition, the target market, the value chain, the revenue mechanism, the value network, and a competitive strategy). Chesbrough (2007) presents business models in stages ranging from type one, undifferentiated business models to type five, business models integrated with innovation processes, and finally to type six, business models as adaptive platforms. That is, while differentiating business models from technology and knowledge, he integrates the innovation of business models with innovations made through technology and knowledge or combines business models with technology and knowledge in the form of a platform for business models of types five and six.

A mediocre technology pursued within a great business model may be more valuable that a great technology exploited via a mediocre business model (Chesbrough 2010). Barriers to business model innovation are the conflict between a business model established for existing technology and that which requires exploiting an emerging, disruptive technology. According to Chesbrough (2010), the successful leadership of organizational change must be brought to bear in order to overcome these barriers. Although business models are closely related to technology, they are an independent corporate element distinguished from technology or knowledge (Chesbrough 2010).

Business models include discussions at the three dimensions of e-commerce, strategy, and technology and innovation management (Zott et al. 2011). Business models in e-commerce describe new gestalts and Internet-based ways of "doing business." Business models focusing on the strategy dimension explain new network and activity system-based value creation mechanisms and sources of a competitive advantage. Business models in relation to technology and innovation management seek to understand how technology is converted into market outcomes. None of these business models from these three viewpoints are for knowledge or technology per se but are instead core elements of a firm's existence and development.

To put the foregoing diverse discussions about business models together, first, business models are concepts materially and theoretically distinguished from technology or knowledge per se. Second, business models are indispensable for the survival and development of firms. Therefore, firms should seek continuous development or the introduction of new business models suitable for leading market environments, referring to the (open) innovation of business models. Third, although business models are not technologies, new knowledge and technologies from various aspects are necessary for the (open) innovation of business models.

Third, we review the relationship between knowledge strategies and business models. Knowledge is considered increasingly as a principal success factor or as the major driving force behind business success (Papavassiliou and Mentzas 2003). Knowledge modeling in weakly structured business processes will be a new approach for integrating knowledge management and business process management. Knowledge-intensive business processes tend to be characterized by dynamic changes of goals, information environments, constraints, and highly individualized and ad hoc communication and collaboration patterns (Papavassiliou and Mentzas 2003).

Business process modeling (BPM) as a tool for knowledge management allows the transformation of informal knowledge into formal knowledge and facilitates its externalization and sharing (Kalpic and Bernus 2006). Business process models focus on descriptions of business process features and characteristics. However, organizations should be aware of the complete knowledge flow process by considering the flow between the organization and the external world and the flows among individuals within (and outside) the organization before undertaking business process modeling (Kalpic and Bernus 2006). All four knowledge processes of internalization, externalization, combination, and socialization can be included as part of the process of business process modeling (Nonaka and Takeuchi 1995).

Different prosperity levels of software firms, as exemplified by the characteristics of size, age, intellectual capital, absorptive capacity, and ownership structures affect their decisions to base their business strategies on an open source software (OSS) supply or on the proprietary distribution of products and services (Harison and Koski 2010). That is, for software companies, corporate strategies for technology and knowledge are very closely related to business processes. In the case of the UK service industry as well, analyses have shown that causal links exist between service firms' knowledge investments, their innovation outputs, and business growth (Love et al. 2011). Love et al. (2011) posit interconnections between knowledge sourcing, knowledge transformation, and knowledge exploitation using innovation value chains and focus on the inflows and outflows of knowledge and technology from or to the outside in each of the connections and processes. That is, they present innovation value chains per se as aspects of open innovation knowledge strategies.

According to a case study of an entrepreneurial venture, the business model is a narrative and calculative device that allows entrepreneurs to explore a market; it plays a performative role by contributing to the construction of the techno-economic network of an innovation (Doganova and Eyquem-Renault 2009). According to a survey on venture capital companies in their early stages, business models serve as brokers to connect firms' technologies and knowledge to markets. According to the pay-as-you-drive (PAYD) auto insurance concept, an incumbent firm profits from business model innovation (Desyllas and Sako 2013). According to the PAYD case, conceptualizing and implementing a novel business model is not sufficient for a firm to achieve a competitive advantage and above average returns. That is, sustained growth and development are not guaranteed by only the (open) innovation of business models.

Value creation by a knowledge-based ecosystem facilitates the innovation process for individual companies and creates innovation communities (Van der Borgh et al. 2012; Yun and Mohan 2012a, b). A knowledge-based ecosystem is a social and economic environment providing specific value sources sought by individual companies in their continuous quest to improve performance. Related to this, the coevolution of the ecosystem's business model with firm-level business models explains why technology-based firms join, stay in, or leave the ecosystem at a certain point in time. That is, the corporate-level business models and knowledge-based ecosystems faced by firms affect each other while also developing and growing together.

The cyclic innovation model (CIM) connects technological capabilities with market needs (Berkhout et al. 2010). It also presents these processes during the process of innovation in the form of a circle of change, including a scientific exploration–technological research–product creation cycle and a scientific exploration–market transitions–product creation cycle. The CIM postulates the coexistence of two cyclical interactions instead of conceptual chains of knowledge. The two cycles mediated by entrepreneurship demonstrate the cyclical coevolution of the introduction of new business models and the introduction of new knowledge and technology.

The business model serves as a mediating concept between technology and economic value (Chesbrough and Rosenbloom 2002). A successful business model creates a heuristic form of logic that connects technical potential with the realization of economic value and constrains the subsequent search for new, alternative models for other technologies later on—an implicit cognitive dimension overlooked in most studies of this topic (Chesbrough and Rosenbloom 2002). This approach considers knowledge in a firm as limited. Because the approach regards the creation of value as useful to consumers through knowledge and technology in a knowledge-based economy. Connection between technology and market as the business model clarifies the meaning of business models as distinguished from knowledge or technology. Discovery-oriented research often produces spillover technologies that lack a clear path to markets; thus, discovering a viable business model for these spillovers is a critical and neglected dimension of creating value from technology (Chesbrough and Rosenbloom 2002). Eventually, open innovation for the discovery and introduction of technology and open innovation for the introduction of a new business model exist in different dimensions.

That is, a common argument of diverse discussions is that although firms' knowledge strategies for the acquisition of technology and business models for the creation of new value are linked to each other in the dynamic process of innovation, they have independent functions and values. Therefore, active open innovation at the corporate level for the acquisition of new knowledge and activities at the corporate level in turn for the acquisition of new business models have different logics and procedures from each other.

9.2.2 Relationship Framework Between Business Model and Open Innovation

The openness of knowledge strategies has been dealt in-depth in different aspects of knowledge management and is regarded as a new core element of corporate strategies in recent strategy studies (Chesbrough 2003; Choo and Johnston 2004; Yun and Ryu 2012; McEvily and Marcus 2005; McNichols 2010; Smith 2001; Xu et al. 2010; Yigitcanlar et al. 2012).

Furthermore, as the importance of not only knowledge or technologies per se but also business models that connect the relevant technologies with markets has come to the fore, actively changing and adding business models as required by customers and markets is appearing as an indispensable element targeted by the new open innovation of SMEs (Berkhout et al. 2010; Chesbrough 2007, 2010; Chesbrough and Rosenbloom 2002; Doganova and Eyquem-Renault 2009; Harison and Koski 2010; Johnson et al. 2008; Kalpic and Bernus 2006; Osterwalder and Pigneur 2002; Papavassiliou and Mentzas 2003; Zott et al. 2011).

Here, open innovation in knowledge strategies consist of cases in which firms actively introduce knowledge and technologies from the outside as well as cases where firms actively release their unutilized technologies outside, although these cases are not common among SMEs. In contrast, closed innovation in knowledge strategies consists of cases where firms establish and implement knowledge strategies centered on their own technologies and knowledge.

Open innovation in business models refers to the active introduction of new business models by firms from the outside (Yun and Cho 2014). When firms have activated open innovation in business models, the addition of new products appear rapidly. In contrast, closed innovation in business models refers to cases where firms maintain their existing products or services, corresponding to business model strategies insensitive to demands from consumers or markets.

The sustainable growth strategies of SMEs will appear differently according to the four types in Fig. 9.1. That is, C-C-type SMEs that use closed innovation approaches in knowledge strategies and during the introduction of business models are assumed to pursue development only within the technologies and business models. They cannot survive for a long time because they are insensitive to consumer

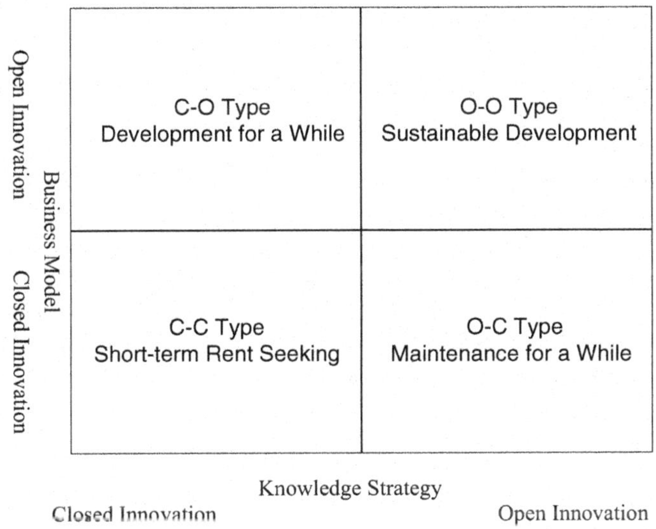

Fig. 9.1 Relationship framework between open innovation and business model (Source: Yun et al. (2015))

preferences, changes in market situations, and the introduction of new knowledge and technologies.

C-O-type SMEs continuously produce new products and services that meet the demands of markets and consumers within their technologies and knowledge. These firms are assumed to develop until they can build diverse business models within their existing knowledge and technologies.

O-C-type SMEs are firms that have secured an area and customers to a certain extent in the market. Therefore, many of them should be considerably medium to large enterprises. These firms can survive until they can supply more efficient products through the introduction of newer knowledge and technologies to their secured market and customers. Therefore, these firms are assumed to be unable to survive if their targeted markets' or customers' demands change rapidly, as changes in their business models are slow; as a result, they cannot sufficiently and actively respond to changes.

O-O-type SMEs not only actively change major products in accordance with customers' or markets' demands but also continuously secure necessary knowledge and technologies from the outside. This type of SME is assumed to be a representative type that can continuously grow and develop. Many technology-based venture companies in countries or regions where knowledge ecosystems are well established and where activities of technology-based venture companies are active are in this category. These firms will grow into global enterprises through sustained growth (Van der Borgh et al. 2012).

9.3 Changing of Open of SMEs

First, in SMEs, there are two different levels of open innovation; therefore, we should bear in mind open innovation in both knowledge strategies and business model in reality. Whereas the former appears as joint research and development with universities—or company—funded research institutes, technology transfers, and research contracts, for instance, the latter appears as new product- or production-related ideas or as feedback and materialization of the demands of consumers and markets for the diversification of existing products. In interviews with firms' chief executive officers (CEOs) or chief technology officers (CTOs), it was identified that SMEs had different approaches with regard to the two factors (Chesbrough 2010). The differences develop into different steps during the process of the development of open innovation by the SME (Fig. 9.2).

SMEs start with a single technology and a single business model and develop into firms with complex technologies and complex business models. O-O-type firms are mostly not very old firms that started as the C-C type and are continuously growing through open innovation in both knowledge strategies and business models. While undergoing the second step of open innovation, these firms are divided into two contrasting open innovation strategy groups: O-C-type firms and C-O-type firms. These two groups are identical in that both have in-house research and development centers and are much more advanced than O-O-type firms in terms of their sizes, ages, and the diversity of their business models. However, the former have larger internal facilities and infrastructures as well as research and development systems than the latter. This focuses on cooperation in technology development and transfers with external parties, reflecting the results in product development and business model additions. On the other hand, the latter more actively implements strategies to cooperate directly with external parties in developing products or adding business models to obtain products or business models from external parties or to complement existing products or business models, thereby having more diversified business models than the former. However, these types of firms have been shown to have in-house facilities and infrastructures or research and development systems which are larger than those of O-O-type firms but generally smaller than those of O-C-type firms. If firms fail in continuing to realize open innovation, they will fall back and become C-C-type firms again, and their sales and firm sizes will undergo a process of stagnation and subsequent shrinking. These firms experience a considerable time of stagnant sales initially before they experience rapid shrinkage. Of course, they may have opportunities for new growth if they switch to complete open innovation approaches in terms of both their knowledge strategies and business models (Cefis and Marsili 2005).

Second, the integrated lengths of the life span, firm size, or sales by SMEs according to the type of open innovation used were determined, as shown in Fig. 9.3. C-C-type firms showed rapid sales growth and larger firm sizes at the beginning but experienced considerable limitations with regard to the magnitude of growth, whereas O-O-type firms grew more slowly compared to C-C-type firms. They grow to higher levels in terms of their life cycles or sizes (Buenstorf 2007). However, a

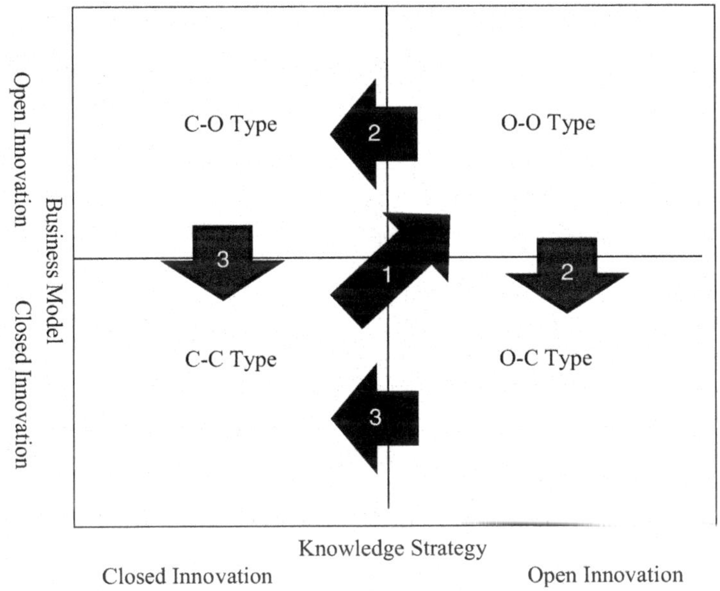

Fig. 9.2 Changing steps of open innovation type in SMEs (Source: Yun et al. (2015))

noteworthy fact in relation to SMEs is that complete open innovation strategies do not guarantee sustained growth. In contrast, it has been shown that they grew further, as did C-O and O-C types, when they secured competitiveness by ensuring their own technological capabilities or through high differentiation in their business model (Lall 1992).

9.4 Conclusion

From this chapter, we find several implications for studies of management and SMEs, as follows. First, SMEs cannot survive if they do not accept open innovation in their knowledge strategies and business models. SMEs that show absolute limitations in resources and manpower should absolutely implement open innovation strategies to secure more diverse resources from markets and external knowledge bases rather than preparing all resources and capabilities by themselves (Van de Vrande et al. 2009; Yun and Mohan 2012a, b).

Second, SMEs should bear in mind the two different levels of open innovation, like knowledge strategies and business models (Chesbrough 2007; Chesbrough and Appleyard 2007). In particular, surrounding the acquisition of diverse business models, SMEs' open innovation strategies appear in more diverse forms (Yun and Mohan 2012a, b). It can be seen that during the process of development, open innovation in business models develops into forms quite different from that of knowledge strategies. Therefore, securing the networks that would enable SMEs to

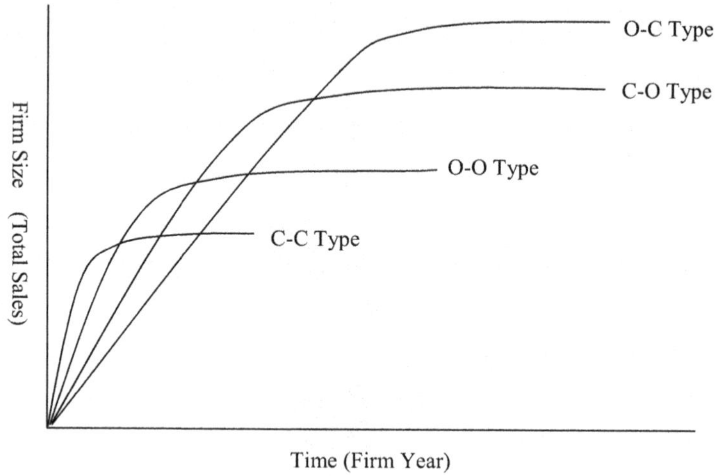

Fig. 9.3 Relationship between the open innovation type and the life cycles and sizes of firms (Source: Yun et al. (2015))

implement different characteristics of an open innovation business model simultaneously by participating in diverse forms of networks in their cluster outweighs anything else (Cooke 2005; Yun et al. 2010).

Third, SMEs should not confuse temporally sustainable development and infinite sustainable development. Firms that use closed innovation strategies in both their knowledge strategies and business models can also grow for some time. However, due to the deepening of the knowledge-based economy, not only is the amount of knowledge existing in the world and the speed of the distribution of knowledge increasing but the demands and expectations of customers also exist in the market immediately through SNS (social networking sites) and similar channels (Yun and Ryu 2012). As a result, cutting-edge products fall into commodity traps more quickly and technology life cycles are shortened (Aveni 2010; Chesbrough 2011). In addition, as more effective technological catch-up strategies have appeared, late movers have been able to chase forerunners in a short time (Kumar and Russell 2002; Lee and Lim 2001). SMEs should take appropriate open innovation portfolio strategies for their knowledge strategies and business models in accordance with their situations.

Fourth, SMEs should choose strategies of open innovation in their knowledge strategies and business models together. According to the choice between open innovation in a knowledge strategy and in a business model, firms have different growth limits and sizes.

It is true that a case study is a method that provides an excellent qualitative analysis in studies of firms (Yin 2008).

Research Question

1. What is the difference of open innovation in knowledge strategy and business model?
2. Find out the examples of open innovation in knowledge strategy and business model in newspaper?

References

Aveni RA (2010) Beating the commodity trap: how to maximize your competitive position and increase your pricing power. Harvard Business School Press, Boston

Berkhout G, Hartmann D, Trott P (2010) Connecting technological capabilities with market needs using a cyclic innovation model. R&D Manag 40(5):474–490

Buenstorf G (2007) Evolution on the shoulders of giants: entrepreneurship and firm survival in the German laser industry. Rev Ind Organ 30(3):179–202

Carrillo FJ, Batra S (2012) Understanding and measurement: perspectives on the evolution of knowledgebased development. Int Journal Knowl Based Dev 3(1):1–16

Carrillo FJ (2015) Knowledge-based development as a new economic culture. J Open Innov Technol Market Complex 1(1):1–17

Cefis E, Marsili O (2005) A matter of life and death: innovation and firm survival. Ind Corp Chang 14(6):1167–1192

Chesbrough H, Rosenbloom RS (2002) The role of the business model in capturing value from innovation: evidence from Xerox Corporation's technology spin-off companies. Ind Corp Chang 11(3):529–555

Chesbrough HW (2003) Open innovation: the new imperative for creating and profiting from technology. Harvard Business Press, Boston

Chesbrough H (2007) Business model innovation: it's not just about technology anymore. Strateg Leadersh 35(6):12–17

Chesbrough HW, Appleyard MM (2007) Open innovation and strategy. Calif Manage Rev 50(1):57–76

Chesbrough H (2010) Business model innovation: opportunities and barriers. Long Range Plann 43(2):354–363

Chesbrough H (2011) The case for open services innovation: the commodity trap. Calif Manage Rev 53(3):5–20

Cooke P (2005) Regionally asymmetric knowledge capabilities and open innovation: exploring 'Globalisation 2'—a new model of industry organisation. Res policy 34(8):1128–1149

Creswell JW (2012) Qualitative inquiry and research design: choosing among five approaches. Sage Publications, Los Angeles

Desyllas P, Sako M (2013) Profiting from business model innovation: evidence from Pay-As-You-Drive auto insurance. Res Policy 42(1):101–116

Doganova L, Eyquem-Renault M (2009) What do business models do?: innovation devices in technology entrepreneurship. Res Policy 38(10):1559–1570

Endres LM, Endres SP, Chowdhury SK, Alam I (2007) Tacit knowledge sharing, self-efficacy theory, and application to the Open Source community. J Knowl Manag 11(3):92–103

Harison E, Koski H (2010) Applying open innovation in business strategies: evidence from Finnish software firms. Res Policy 39(3):351–359

Johnson MW, Christensen CM, Kagermann H (2008) Reinventing your business model. Harv Bus Rev 86(12):57–68

Kalpic B, Bernus P (2006) Business process modeling through the knowledge management perspective. J Knowl Manag 10(3):40–56

Kumar S, Russell RR (2002) Technological change, technological catch-up, and capital deepening: relative contributions to growth and convergence. Am Econ Rev 92(3):527–548

Lall S (1992) Technological capabilities and industrialization. World Dev 20(2):165–186

Lee K, Lim C (2001) Technological regimes, catching-up and leapfrogging: findings from the Korean industries. Res Policy 30(3):459–483

Lopez VWB, Esteves J (2012) Acquiring external knowledge to avoid wheel re-invention. J Knowl Manag 17(1):6

Love JH, Roper S, Bryson JR (2011) Openness, knowledge, innovation and growth in UK business services. Res Policy 40(10):1438–1452

McEvily B, Marcus A (2005) Embedded ties and the acquisition of competitive capabilities. Strateg Manag J 26(11):1033–1055

McNichols D (2010) Optimal knowledge transfer methods: a Generation X perspective. J Knowl Manag 14(1):24–37

Morris M, Schindehutte M, Allen J (2005) The entrepreneur's business model: toward a unified perspective. J Bus Res 58(6):726–735

Nelson RR, Winter SG (1982) An evolutionary theory of economic change. Belknap P of Harvard Up, Cambridge

Nonaka I, Takeuchi H (1995) The knowledge-creating company: how Japanese companies create the dynamics of innovation. Oxford University Press, New York

Ordonez de Pablos P (2002) Evidence of intellectual capital measurement from Asia, Europe and the Middle East. J Intellect Cap 3(3):287–302

Osterwalder A, Pigneur Y (2002) An eBusiness model ontology for modeling eBusiness. BLED 2002 Proceedings, p 2.

Papavassiliou G, Mentzas G (2003) Knowledge modelling in weakly-structured business processes. J Knowl Manag 7(2):18–33

Peters M, Schneider M, Griesshaber T, Hoffmann VH (2012) The impact of technology-push and demand-pull policies on technical change—does the locus of policies matter? Res Policy 41(8):1296–1308

Seidman I (2006) Interviewing as qualitative research: a guide for researchers in education and the social sciences. Teachers College Press, New York

Smith EA (2001) The role of tacit and explicit knowledge in the workplace. J Knowl Manag 5(4):311–321

Timmers P (1998) Business models for electronic markets. Electron Mark 8(2):3–8

Van de Vrande V, De Jong JP, Vanhaverbeke W, De Rochemont M (2009) Open innovation in SMEs: trends, motives and management challenges. Technovation 29(6):423–437

Van der Borgh M, Cloodt M, Romme AGL (2012) Value creation by knowledge-based ecosystems: evidence from a field study. R&D Manag 42(2):150–169

Wenger E (2004) Knowledge management as a doughnut: shaping your knowledge strategy through communities of practice. Ivey Bus J 68(3):1–8

Wei Choo C, Johnston R (2004) Innovation in the knowing organization: a case study of an e-commerce initiative. J Knowl Manag 8(6):77–92

Xu J, Houssin R, Caillaud E, Gardoni M (2010) Macro process of knowledge management for continuous innovation. J Knowl Manag 14(4):573–591

Yigitcanlar T, Metaxiotis K, Carrillo FJ (2012) Building prosperous knowledge cities: policies plans and metrics. Edward Elgar Publishing, Cheltenham

Yin RK (2008) Case study research: design and methods. SAGE Publications Incorporated, Haidelberg

Yun JHJ, Cho BJ (2014) An exploratory study of the economic effect of open innovation. J Sci Technol Policy Manage 5(1):24–40

Yun JHJ, Park S, Avvari MV (2011) Development and social diffusion of technological innovation cases based on mobile telecommunications in national emergency management. Sci Technol Soc 16(2):215–234

Yun JHJ, Mohan AV (2012a) Exploring open innovation approaches adopted by small and medium firms in emerging/growth industries: case studies from Daegu–Gyeongbuk region of South Korea. Int J Technol Policy Manage 12(1):1–19

Yun JHJ, Mohan AV (2012b) The relation between internal and external open innovation: a study of firms located in the Goomi and Banwol-Sihwa clusters in South Korea. In: Brem A, Tidd J (eds) Perspectives on supplier innovation: theories, concepts and empirical insights on open innovation and the integration of suppliers. Imperial College Press, London, pp 257–286

Yun JHJ, Ryu G (2012) A study on the difference of open innovation effect between modular and non-modular firms in Korea. Asia Pac J Innov Entrepren 6(1):56–72

Yun JHJ, Park S, Lim DW, Hahm SD (2010) Emergence of East Asian TFT-LCD clusters: a comparative analysis of the Samsung cluster in South Korea and the Chimei cluster in Taiwan. Asian J Technol Innov 18(1):201–228

Yun JHJ, Jung WY, Yang JH (2015) Knowledge strategy and business model conditions for sustainable growth of SMEs. J Sci Technol Policy Manage 6(3):246–262

Zott C, Amit R, Massa L (2011) The business model: recent developments and future research. J Manag 37(4):1019–1042

Part V

Developing a Business Model

The Way from Open Innovation to Business Model

10

Abstract

This chapter has five goals. The first of these is to look into the relationship between open innovation (OI) and business models (BMs). Second, it seeks to analyze business models as a combination of technology and the market. Third, it will explain the four perspectives of business models from OI knowledge funnel. The fourth is to understand the five key factors of business models from "the principle of the Five Ws and One H." Fifth and last, it seeks to discuss good habits for developing the capabilities of business models.

Keywords

Open innovation • Business model • New combination

10.1 The Relationship Between Open Innovation and Business Models

Open innovation (OI) occurs as an open and creative connection between technology and the market. In particular, if a person who develops a technology differs from one who uses the technology to produce a product or service and supplies it to the market, the situation is termed *open innovation*. Open innovation is a concept with a firm-based definition. That is, a firm introduces an external technology and innovates with the existing product; it produces a new product or creates a new market or transfers a technology that is not in use to another company to produce a product or create a new market.

However, a business model (BM) is created through a new combination of technology and the market. When this technology goes beyond the boundary of the

J.J. Yun, *Business Model Design Compass*, Management for Professionals,
DOI 10.1007/978-981-10-4128-0_10

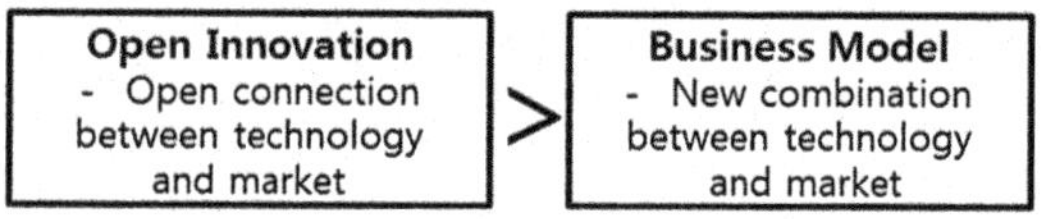

Fig. 10.1 The size relationship between open innovation and BMs

company and advances to open innovation, the technology is newly combined with a market (Yun et al. 2016). The end product is referred to as a BM.

From the perspective of a company, open innovation is larger than a BM, as shown in Fig. 10.1. Figure 10.1 shows that open innovation is conceptually larger than BMs in terms of size. That is, open innovation and BMs share the same concept qualitatively. However, the category of open innovation is significantly larger than that of BMs.

In terms of content, a BM is largely subject to open innovation, as shown in Fig. 10.2. Section A only has an open connection between technology and the market without a creative combination. Section B, on the other hand, has a creative combination of technology and the market. Lastly, Section C, which is only a small portion of the BM development of a company, carries a part of the combination of technology and the market through the closed innovation (CI) of a company.

However, Section C moves toward open innovation when the BM is implemented in a market, as shown in Fig. 10.3. That is, it moves to Section B and becomes a target of open innovation. More specifically, the rate at which a BM is created in CI is low, with its implementation period having only a limited duration. The reason behind this is that the volume of the knowledge of the world skyrockets, and the distribution speed significantly rises owing to the advent of the mobile Internet era. In addition, as the connected economy emerges, the global economy and the actual market concretely connect. Thus, the time when a specific BM exists in an isolated market is significantly limited.

The qualitative relationship between open innovation and a BM is described in Fig. 10.4. That is, the open BM has an open connection between the technology and the market of open innovation and a creative combination of technology and the market (Yun 2015; Yun et al. 2015). However, taking into consideration the fact that the CI BM is diminutive and is a phenomenon that exists temporarily, most BMs are included in an open BM.

10.2 Business Models as New Combinations Between Technology and the Market

10.2.1 Technology Aspects of Business Models

Technology has three aspects, as shown in Fig. 10.5. The protected knowledge of Section A, which is a target of intellectual property, receives the most attention from the research and development arms of companies and the government. However, for the technology of a BM, Section B, *Protectable Knowledge*, protected by companies for their security, and Section C, *Ideas or Knowledge*, well known to the general public, carry significance.

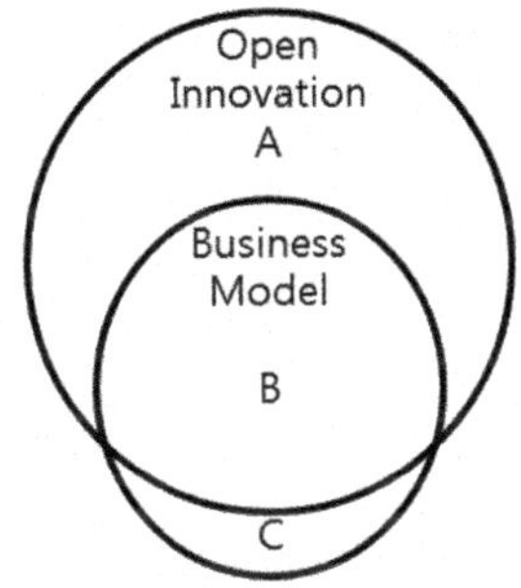

Fig. 10.2 The content relationship between open innovation and BMs

A common case related to Section B is an instance in which a company whose corporate security is a method to produce Coca-Cola continuously develops various BMs beyond its sales of finished beverages. In addition, Elon Musk of Tesla Motors and SpaceX retain technologies as corporate secrets. They also manufacture various electric cars, including high-end automobiles; produce battery packs and various rockets; and create additional BMs based on core technology. This situation is also an example of Section B.

Above all, it is necessary to concentrate on Section C, in which technology is a target for the development of a BM and is the largest source. The escrow BM patent of PayPal, a key payment system in e-commerce, and the one-click patent of Amazon, which makes the company a global power in various e-commerce markets, are based on simple ideas rather than an original patent.

However, the most important technical aspect of a BM is, if the BM patent based on the technology of Section C additionally secures the technology of Section A or B, that the sustainability and the competitiveness of the BM can be significantly improved. In contrast, for the Section A technology patent, developing a BM based on a patent or combining a BM based on the technology of Section C can achieve technology commercialization with higher creativity and success.

10.2.2 Market Types of Business Models

There are three types of markets that become a target of a BM, as shown in Fig. 10.6: a modern market, a potential market, and a social market. These three items should be given consideration when a new BM is created through a combination of technology and the market. In the case of the modern market, it is necessary to examine a market that exists in another space. For example, a common sports beverage market in Southeast Asia is seen as refined and exists as a modern market. In addition, the existing market of a different industry is important to consider for the development of a new BM. Examples are the car-sharing business market of Uber, with a BM that differs from that of the original taxi market system, and the new reservation service BM of Hotels.com or Booking.com, going beyond what is already available in the current hotel industry.

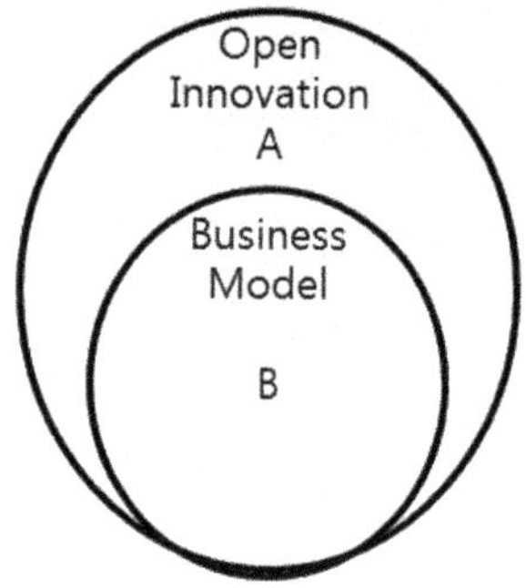

Fig. 10.3 Disappearance of a BM in CI

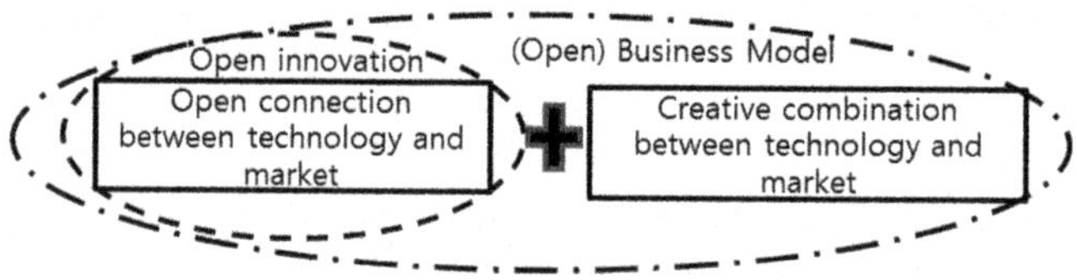

Fig. 10.4 Quality relationship between open innovation and a BM

The potential market, on the other hand, is not active now but is expected to significantly grow in the future. At present, space trips, personal intelligent robots, perfect autonomous vehicles, and personal medical system markets are potential markets that are expected to grow considerably later. The potential market is considered with priority when a new BM is developed. However, this market takes a long time to develop, and its uncertainty is significant. Thus, a BM combined with a potential market needs a very long-term perspective given its extreme uncertainty. In particular, a potential market corresponds to the failure of the conventional market, where the government intentionally becomes involved to develop the market as a future industry.

There are goods and services which are socially needed but are impossible completely to find only through the market function. These items make up the social market, which is a major target of social enterprises. In many cases, the market becomes a public service target through the government. However, all social markets represent a market that leads to the creation of market value at present or in the future, beyond simply creating social value. Most social fields, such as new and renewable energy, services for the socially disadvantaged, the local community, and social activities to eradicate poverty are targets of the social market. This social market becomes an infinite source of potential markets and current markets. As an example, the space development project of the American government becomes the source of the current space industry of the US social market, which supplies Burro batteries to Africa and which is slowly developing as a potential market (Backs 2013, p. 23).

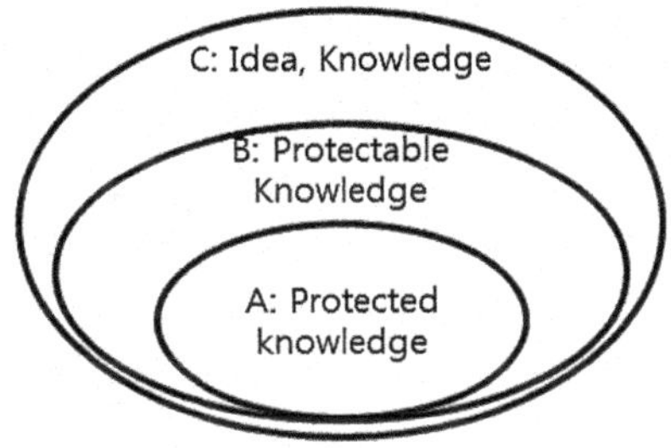

Fig. 10.5 Three aspects of technology (Source: Chesbrough (2003, p. 157))

10.2.3 Quadrants of the Combination of Technology and the Market

The combination of technology and the market has a number of quadrants, as shown in Fig. 10.7. Section A has the largest number of targets for *existing technology* and *existing markets*. Following Section A are Section C, which is based on new technology, and Section B, which is based on new markets. Lastly, Section D has the smallest portion based on new technology and the market.

That is, a BM is created with A > C or D > D. Thus, it is necessary to continue to pay close attention to the development of BMs through new combinations of the three types of markets or technologies and markets in *existing technologies* and even *existing markets*. In fact, in most cases, although many creative technologies are created, they are not combined with markets, which ultimately lead to the failure to develop a new BM.

10.3 Four Perspectives of Business Models from the Open Innovation Knowledge Funnel

Open innovation occurs in the four points of a company's knowledge funnel. *A* in Fig. 10.8, i.e., where open innovation occurs in the research stage of a company, is the starting point. From the perspective of an engineer (i.e., a researcher), when one thinks of a new combination beyond the open connection between technology and the market, it will be the engineer's open innovation-based new BM. This BM is formed by contemplating a new combination of technology and the market beyond the boundary of a company from the point of view of an engineer. For example, there are the electric cars of Elon Musk and the new search engine created by Sergey Brin and Larry Page named *Alphabet*.

Next, the user of the existing firm's open innovation-based BM in category *B* makes a new combination beyond the open connection between technology and the market based on the various requirements and expectations that exist in the supply chain of a company. For example, dental clinics make a BM that produces implants based on various requirements and expectations that arise as they run their medical businesses. In addition, many automotive parts (bicycle) companies in Korea handle the demand for new automotive parts from producers of bicycle parts and textile

Fig. 10.6 Three types of markets

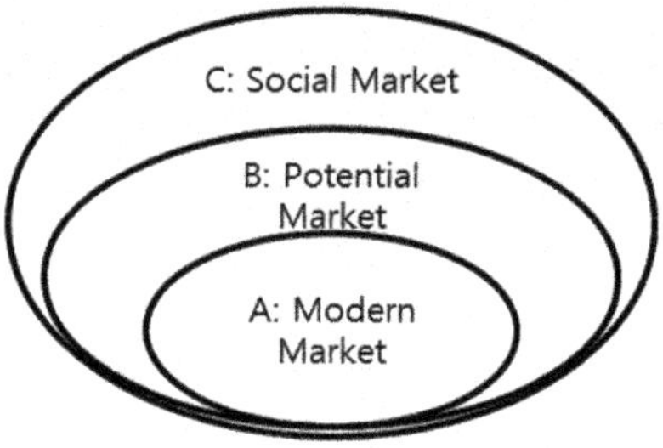

Fig. 10.7 Quadrants of technology and the market

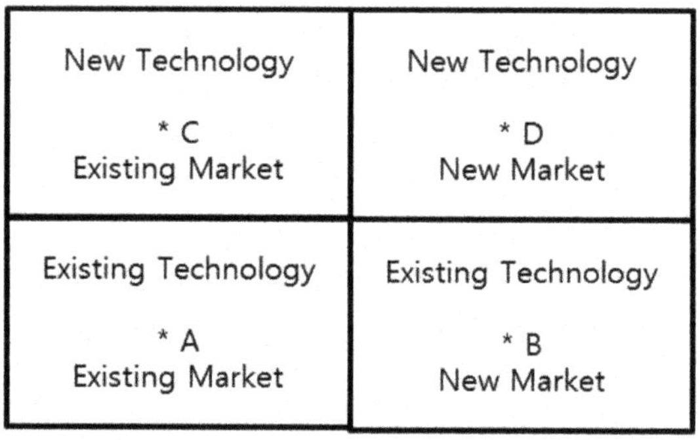

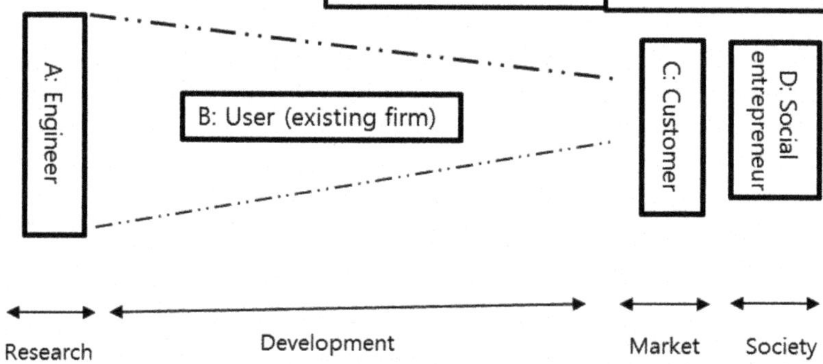

Fig. 10.8 Four points in the open innovation knowledge funnel

machinery; these are then transferred to automotive parts suppliers. Many small- and medium-sized companies around the world respond to the requirements and expectations of a supply chain and make strenuous efforts to develop new BMs. However, it should be noted that the development of a user BM essentially requires open innovation. If companies are not open to new technology and markets, the possibility of developing new BMs will decrease.

Third, the customer open innovation-based BM of *C* is a case in which the leading user of a product starts from his/her consumption experience and achieves a creative combination of technology and the market. *C* is a BM development point that creatively combines technology and the market based on the personal experience of a customer while going beyond the technology of a potential company. Most lean start-ups are based on this customer open innovation-based BM. The core of customer creation indicates *C*, which starts from the critical mind of a consumer and, from there, paves the way for the creation of a new BM. Prime examples of this are how Dyson vacuum cleaners started from the critical mind of Dyson's founder and how Han Kyung-Hee cleaners took off from simple customer requirements of wiping and dust absorption.

Fourth, this BM is created based on the creative combination of technology and the market from the perspective of social necessity, in which the market does not currently exist concretely beyond a corporate market. This BM is formed through a creative combination of the social market and technology beyond the potential market of a company, giving it the features of freedom and emergence. In fact, for this BM in *D*, the characteristics and capacity of a BM from *C* to *A* or *B* are expanding, in general, owing to the category relativity of the market and society and the knowledge capitalization created by the acceleration of the knowledge-based economy.

10.4 Five Key Factors of Business Models

A BM can be more clearly identified by presenting core components. There are various ways in which components are presented in a BM. Some examples are (a) in three core components (resources and competences, organizational structure, and its propositions for value delivery; Demil and Lecocq 2010), (b) the four-box BM framework (customer value proposition, profit formula, key processes, and key resources; Johnson 2010 p. 45), (c) and the Business Model Canvas (BMC) with nine elements (customer segments, value propositions, channels, customer relationships, revenue streams, key activities, key resources, key partnerships, and cost structure; Osterwalder and Pigneur 2010, p. 15).

This research organizes a BM into five elements, with no additional elements required owing to the extreme level of simplicity. This method will also allow easier composition of a BM, as not too many elements must be taken into consideration.

Another reason behind this limited number of elements is that the principle of the Five Ws and One H is the concept most commonly used by society. These five BM components directly apply the said principle to a BM in a logical manner. This advantage allows the possibility of including the elements of earlier work.

Similar to Table 10.1, the five elements based on the principle of Five Ws and One H are the core parts of a BM. First are the concrete *customer segments*, wherein, for customers, defining to create a highly concrete and visible *persona* is essential to set all of the BM elements in the future. In fact, whether customers are concretely and successfully set is the starting point that determines the success of the development of a BM. For the customer setting, with the method of exploring the area without customers, that is, the *noncustomer* area (the adjacent market customers as targeted new customers adjacent to the current market), together with the overshooting customers who exceed the requirements and the expectations of existing customers, new customer groups are found.

Second are the concrete *value propositions* such as problem resolution or desire satisfaction. Values cannot be provided to customers in a completed form at one

Table 10.1 Five key factors of BMs

Five factors	Contents	Main considerations
Customer segmentation (who)	Concretely define a customer	Persona Noncustomer Adjacent market Overshooting
Value propositions (what)	Resolve a problem Provide a concrete value Meet a desire	Minimum viable product (MVP) Feedback Pivot
Technological system (how)	Concrete system that creates value by combining technology and the market	Key resources Key activities Key partnerships
Revenue and cost (why)	Cost structure and contents Profit structure and contents	Revenue streams Cost structure
Customer meeting (when and where)	How to meet a customer How to form customer relationships	Channels Customer relationships

Source: Osterwalder and Pigneur (2010), Aulet (2013), Blank (2013)

time. By creating a minimum viable product (MVP) to provide to customers as rapidly as possible and thus meet customer requirements and expectations that can be additionally checked through feedback, the contents and quality of *value propositions* should be developed. This indicates that when one gives a product to a customer, the important value that should be considered is meeting the time requirements of the customer even by providing a *pivot* in addition to good product quality. For example, Microsoft Corporation provided Windows OS from its beta version with the priority to secure customers; the company then improved the system continuously according to the requirements and expectations of the customers.

Third is the *technological system* that combines technology and the market and creates added value. It corresponds to a point that meets customers by systemically combining key resources, key activities, and key partnerships. With this, the success of the patent of a BM depends on the novelty and creativity of the *technological system*. According to the international patent standard, a BM patent can be registered if it has a systemic combination of IT-based technology and the market. Thus, if the *technological system* of a patent combines central components based on IT creatively, the possibility of specialization increases.

The fourth element consists of the structure and contents of *revenue and cost* that form a BM. The sustainability of a BM depends on how revenue and cost can be creatively and simply implemented for a short period. A sustainable BM design is determined by the possibility of realizing the structure and contents and the balance of the long-/short-term distribution. With an achievable structure of *revenue and cost*, if their short-/mid-/long-term contents are balanced, the BM is considered to be sustainable.

Fifth and last is how we meet a customer and form a relationship with a customer, termed a *customer meeting*. This is the final element to consider when gauging the success of a BM. It cannot be ignored that the creativity and sustainability of the method and structure of meeting a customer and forming a relationship with the customer are, ultimately, highly significant steps to secure the patent of a BM.

10.5 Good Habits for Developing Capabilities of Creative Business Models

10.5.1 Walking and Meditation

Having a short meditation period of about 5 min every day is a helpful activity to help develop ideas with regard to the open connection and creative combination of technology and the market beyond the open connection and combination of familiar markets and technology.

Meditation or mindfulness is a method of simply rebooting one's thoughts. By sitting cross-legged on the ground or sitting in a desk, closing one's eyes, concentrating on one's breathing, and emptying one's mind for approximately 5 min, it is easier to contemplate a new BM, a combination of technology and the market, from a completely different perspective. The intense activity of recalling the open innovation and creativity of a BM between different technologies and the market make up the meditation experience itself. It is an important habit that promotes the development of a new and creative open innovation and a BM.

Another helpful activity is taking a casual stroll for more than 30 min, in particular without any dialogue, every day, which has the same effect as meditation. It gives an opportunity to empty one's mind and look into oneself in greater depth. Walking in the sun and feeling nature's bliss in the sky, the earth, trees, flowers, grass, and the wind will help one generate fresh ideas. At the same time, walking while listening to good music as described in Table 10.2 may raise the effectiveness of meditation.

10.5.2 Taking a Trip

Going on a trip is the easiest way to find new markets and technologies. It is recommended to take a trip to urban areas of underdeveloped countries rather than to nations with mature forms of capitalism. When one takes a walk in urban areas, it is easier to form ideas with regard to BMs that meet society's desires. Aside from this, the excitement of these experiences provides a picture of the open connection and creative combination of technology and the market. Examples of these experiences are seeing unrefined articles being sold in the market, experiencing the service provided by coffee shops, seeing buildings whose structure and usage are unsystematically decided, seeing unsophisticated advertisements and expressions, and experiencing nature as a vital part of our lives in an unrefined manner.

With the development of a capitalistic economic system that connects the world and the significant growth of the Internet, a trip to an urban area of an advanced nation is useful preliminarily to verify one's BM being formulated rather than

Table 10.2 Singers or music for walking meditation (examples)

ABBA
Adele
Amy Winehouse
Ariana Grande
Beatles
Billy Joel
Bob Dylan
Bruno Mars
Charlie Parker
David Bowie
Jim Morrison and Doors
Jimi Hendrix
Louis Armstrong
Kenny G
Nana Mouskouri
Seong-Jin Cho
Stevie Wonder
Sting
U2
Wham
Yo-Yo Ma
Arthur Rubinstein (Nocturnes, Op.)
Glenn Gould's piano performances
Bach: Cello Suites Nos. 1–6

finding a new BM. Positive or negative effects that may arise when one's BM is implemented and secured with regard to its sustainability can be verified preliminarily through various experiences during the trip. Taking a stroll along an alley allows the gathering of various experiences, and traveling to advanced nations turns these experiences into actions.

Of course, a meditation trip to the Himalayas, Phnom Penh, Xianggelila, and certain Indian cities will provide more opportunities to organize one's thoughts as opposed to simply taking a walk. Such a trip also allows one to devise the combination of technology and the market in everyday life in a new way.

10.5.3 Enjoying Movies and Reading Books

Watching a movie or reading has the effect of emptying one's mind and allowing one indirectly to gain experience with the connection and combination of technology and the market with a new structure and contents. This indirect experience provides an opportunity to meet a concrete change of a new connection and combination of technology and the market. For example, the movie *Transcendence* vs. *Her* gives an idea of a future market from opposing points of view as predicted when artificial intelligence is developed. Here, one can think about new BMs at various levels to expand reasonable and emotional capacities when a new BM is developed in the same field. In addition, the movie *Island* provides an opportunity newly to access a new BM and open innovation related to the security and safety to be taken into consideration when thinking of a new BM as biotechnology and human DNA storage technology are developed.

The novel *Brave New World* by Aldous Huxley presents an astounding introspection and ideas of interests about the future society with biotechnology already developed before its basic technology takes strong root in the field. *Brave New World* allows the development of open innovation and a creative BM, ahead of time, from several years to tens of years, of the existing market in terms of a creative connection and a combination of technology and the market. With this, a small change attracted by the minority may create *micro trends*, which are significant worldwide. An example of this is *Black Swan*—a meaningful existence of the extreme exception, as described by the statistician Nassim Nicholas Taleb. Through movies, by experiencing a hypothetical black swan or micro trends, one can form a motive with regard to the connection between new and open technology and the market and the combination of creative technology and the market (Taleb 2010, p. 21).

Research Question

1. Select any case firm from new paper, and explain five factors of business model.
2. Select any case firm from new paper, and explain three different markets and technologies of the firm.

References

Aulet B (2013) Disciplined entrepreneurship: 24 steps to a successful startup. Wiley, Hoboken

Backs S M (2013) Bright Lights, No City: An African Adventure on Bad Roads with a Brother and a Very Weird Business Plan: Elsevier Science BV PO BOX 211, 1000 AE Amsterdam, Netherlands.

Blank S (2013) The four steps to the epiphany. K&S Ranch, Pescadero

Chesbrough HW (2003) Open innovation: the new imperative for creating and profiting from technology. Harvard Business School Press, Boston

Demil B, Lecocq X (2010) Business model evolution: in search of dynamic consistency. Long Range Plan 43(2):227–246

Johnson MW (2010) Seizing the white space: business model innovation for growth and renewal. Harvard Business Press, Boston. OpenURL

Osterwalder A, Pigneur Y (2010) Business model generation: a handbook for visionaries, game changers, and challengers. John Wiley & Sons, Hoboken

Taleb NN (2010) The black swan: the impact of the highly improbable fragility, vol 2. Random House, New York

Yun JJ (2015) How do we conquer the growth limits of capitalism? Schumpeterian dynamics of open innovation. J Open Innov Technol Market Complex 1(1):1–20
Yun JJ, Jeong E, Yang J (2015) Open innovation of knowledge cities. J Open Innov Technol Market Complex 1(16):1–20
Yun JJ, Won DK, Park KB (2016) Dynamics of from open innovation to evolutionary change. J Open Innov Technol Market Complex 2(7):1–22

Customer Open Innovation-Based Business Model Developing Circle

11

Abstract

This chapter looks into the relationships among open innovation (OI), entrepreneurship, and the introduction of new business models (BMs) based on case studies of information technology firms at Daegu and Seoul in Korea and in Indonesia. Subsequently, this chapter will develop a perspective on the relationship between open innovation and business model even when they are not generated in the two different contexts of different regional innovation systems (RISs) and different national innovation systems (NISs).

Keywords

Open innovation • Entrepreneurship • New business model • National innovation system • Regional innovation system

11.1 Cases

11.1.1 Dyson Vacuum Cleaners

In 1978, regular vacuum cleaner user James Dyson noticed the poor performance of his device. Thus, he disassembled the item and found that its dust bag was full to the point of decreasing its air suction force, ultimately affecting the cleaner's overall functionality.

Dyson thought that it is "inconvenient for people to replace the dust bag once it gets filled with the dust sucked in by the vacuum cleaner." He added, "Everyone has complaints about a product that does not function properly. As a design engineer, we need to handle these kinds of problems. We focused on product development and improvement." Dyson then began his efforts to develop a vacuum cleaner

J.J. Yun, *Business Model Design Compass*, Management for Professionals,
DOI 10.1007/978-981-10-4128-0_11

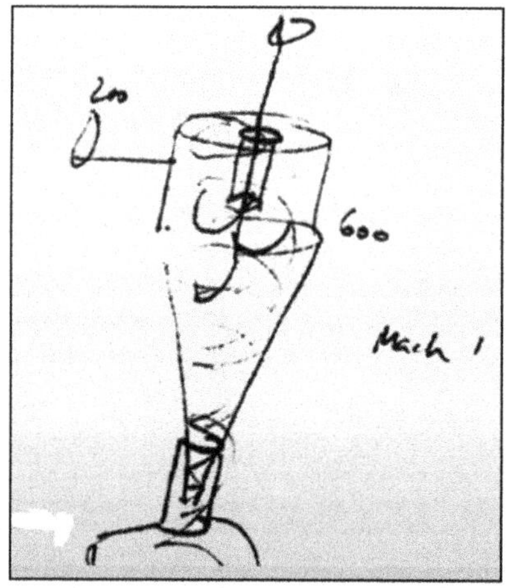

Fig. 11.1 Model image of a Dyson vacuum cleaner without a dust bag (Source: http://www.dyson.com)

without a dust bag, for which he drew the concept map shown in Fig. 11.1. For this endeavor, he applied his experience in working at an industrial cyclone tower, which separated paint particles from air through centrifugal force. He succeeded in achieving his goal after making 5127 prototypes over a course of 5 years.

The first factor to take into consideration in Dyson's case is his primary concern to provide a solution to the inconvenience he, being a vacuum cleaner consumer himself, is experiencing with regard to home appliances. Through this mindset, he eventually became the owner of a global vacuum cleaner company. The second factor is his introduction of technology applied to industrial cyclone towers, that is, a technology that was already in use, to solve his present problem. Third, he applied this technology to create a new concept of a vacuum cleaner. During this application process, he created 5127 prototypes, and through this process, he was able to accumulate the knowledge and technology that became the source of the Dyson product line, which includes air multipliers and air purifiers. Fourth, his products did not compete with the existing mature vacuum cleaner market but created a premium vacuum cleaner market with high functions at a high price, as he made independent off-line products. Fifth, by being defined as an innovative icon rather than merely as an electronic goods company, in addition to its stylish design, Dyson was able to recognize the requirements of both the seller and consumer in the creation of its products.

11.1.2 Han Kyung Hee Steam Cleaner

Kyung-Hee Han, the inventor of the Han Kyung Hee steam cleaner and CEO of the HAAN Corporation, graduated from the Master of Business Administration program at California State University and went to work for the Korean government. At that time, owing to the fact that nearly everyone's daily activities included cleaning, she began to form the idea of creating a steam cleaner to make the cleaning

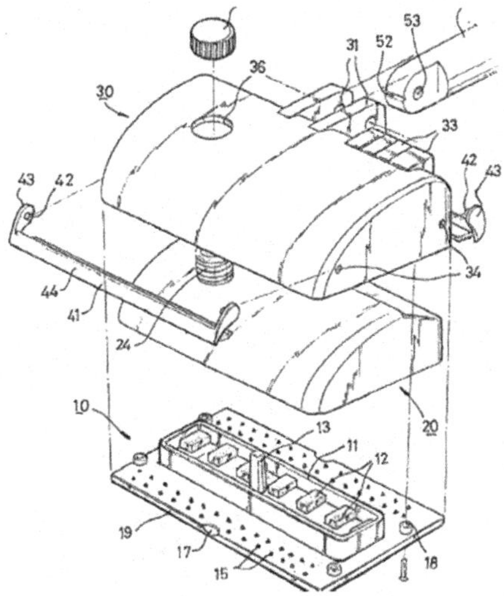

Fig. 11.2 Concept of the Han Kyung Hee steam cleaner (Source: Korea Patent 10-1999-0043562)

process easier. Vacuum cleaners were already widely available but were only customized for the lives of people in the west. She came to the conclusion that Koreans tended to wipe the floor/room after using a vacuum cleaner. She focused on this action while taking into consideration the inconvenience felt by users. She also observed this firsthand when she cleaned using a vacuum cleaner, thus allowing her to take into consideration the experiences of other working women like her. When wiping an area that has already been vacuumed, she noticed that the dirt on the damp cloth mainly spreads with each wipe. This was what she aimed to prevent as she started her efforts to create a new type of cleaner (Cho and Kim 2014). Han concentrated on the inconvenience felt especially by working women including herself and came up with an idea that provided a solution. She then applied for a patent for her idea and established a company to develop and sell related products.

Having a strong critical mind as a customer eventually translates into an invention. This was the case for Han, who also majored in French Language and Literature at Ewha Womans University and graduated from the Ewha School of Business. She applied for a patent with the idea of the steam cleaner described in Fig. 11.2 in October of 1999. The patent Han applied for concerned the continuous generation of hot steam with the use of electrodes, maintaining a proper humidity level with sterilization and evaporating water with smooth cleaning. She built a company in September of 1999 and, since then, has aggressively worked on the development of steam cleaners. The company was able to achieve KRW 100 billion in sales in 2005.

The success factors of the Han Kyung Hee steam cleaner are as follows. First, she paid attention to the problems she, herself, was facing. As the number of working women continues to grow, she acknowledged that time is important, taking note of the demand for cleaners with the functions of both a vacuum cleaner and a wiper. She identified that there is a growing need among working women in Korea for cleaners that significantly decrease their cleaning time and increase the level of cleaning efficiency. Second, to integrate the two processes directly, using a vacuum cleaner and wiping the floors in the rooms, she created a vacuum cleaner that sprays steam. Here, she put the phrase "Necessity is the mother of invention" into action. Third, she configured a system that integrates the cleaning process by combining the functions of a vacuum cleaner and a steam cleaner into a single system. In fact, many products produced by HAAN Corporation are based on a single-system technology that combines vacuuming and steam cleaning. Fourth, she established a lower price for her steam cleaner as compared to the prices of existing vacuum cleaners, and she targeted a niche market. As a new and small player in the existing mature vacuum cleaner market, where large market leaders already existed, she pursued the strategy of entering a niche market, targeting double-income families and unmarried individuals, rather than creating a new market.

11.2 Customer Open Innovation-Based Business Model Developing Circle

As shown in Fig. 11.3, if a customer creatively combines technology and the market by circulating the principle of the Five Ws and One H, a new business model (BM) suggested by the customer is created. What makes this business model special is the fact that it is developed based on the clear perspective of consumers or demand articulation, taking into consideration their status as the lead users or as normal consumers (Kodama and Shibata 2015).

With this, having a mindset in which any company, any engineer who has outstanding skills and knowledge, and any social entrepreneur is a consumer is necessary. That is, a business model developed based on consumer ideas should be given focus, as it independently becomes the source of a creative combination of technology and the market. This business model can also be used as an auxiliary tool to develop other business models of existing firms, engineers, or social entrepreneurs. Thus, it should be noted that the interest in the creative connections and combinations of technologies and the market based on the interests of business model developers is the source of the development of all creative business models.

11.2.1 Who Is in Addition?

11.2.1.1 Key Points

Creative Performance of Interested Consumers

Lead users or normal consumers have sufficient technology and knowledge in the field. They are constantly looking for new alternatives with high creativity with regard to product consumption, and, above all, are innovators in the field.

However, most of them directly lead creative endeavors rather than merely staying and providing feedback, as it is much easier to combine technology and the market creatively, as they exist in the world, and this combination can be realized at a low price (Fig. 11.4).

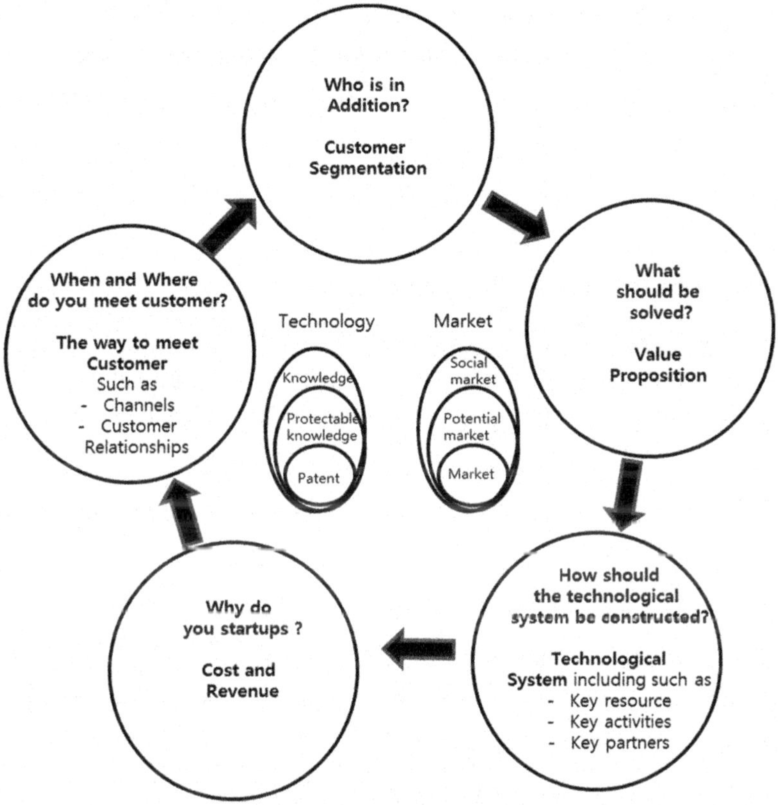

Fig. 11.3 Customer open innovation-based business model developing circle

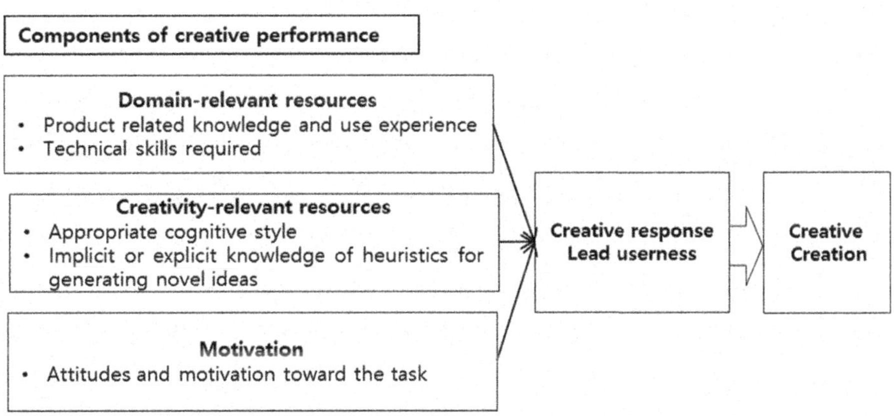

Fig. 11.4 Creativity of lead userness (Source: Praceus (2014, p. 54) modified)

Table 11.1 The philosophy of customer segmentation

The customer can let you know a list of all the features customers want before they buy your product
The customer is actively searching for a solution and has a timetable for finding it
The customer has a committed, or can quickly acquire, budget dollars to solve the problem

Source: Blank (2013, pp. 44 and 46) The customer discovery philosophy, partially extracted and modified

The Philosophy of Customer Segmentation

As presented in Table 11.1, customers have sufficient knowledge of the actual problems being experienced, and, here, some would go out of their way to fix the problems, even investing certain amounts of time and money to complete the task. Thus, having a good understanding of customer groups with a critical mind represents the means of establishing the concrete customers of a new business model.

11.2.1.2 Customer Segmentation for the Customer Open Innovation Business Model Circle

First, it is necessary for a hopeful business model developer to engage in self-definition but from the perspective of a customer. Next, it is necessary to define a customer as an objective third party. Finally, a concrete case should result or a persona should be expressed in a form that can be measured or recognized.

Initially, it is better to have customer segmentation concretely and directly set on a small scale to the greatest extent possible. This segmentation can be expanded, if necessary, after practically or ideologically circulating the customer open innovation business model circle and concretely creating and polishing a business model.

Setting an extremely wide customer segmentation from the beginning mainly causes the customer open innovation business model circle to be stagnant, preventing its progress to the next stage.

Customer Segmentation Template: Customer Open Innovation Business Model Circle

1. Describe your role as the person experiencing the problem.

2. Describe objective customers who are experiencing this problem.

3. Describe a concrete customer group through a measurable and recognizable form, for example, by showing practical cases or creating personas.

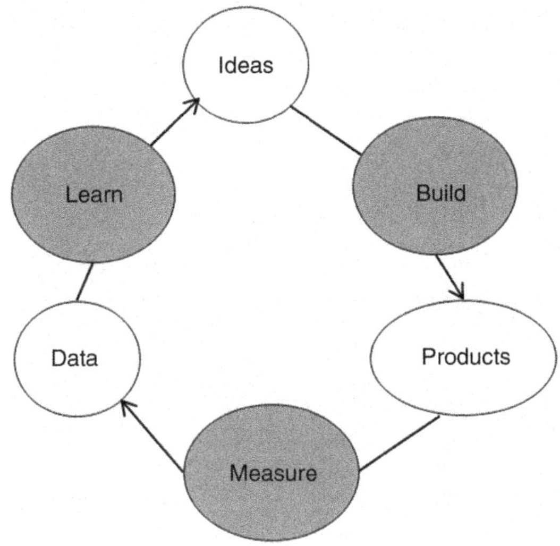

Fig. 11.5 Build–measure–learn loop (Source: Maurya (2012, p. 12), Reis (2011, p. 75))

11.2.2 What Should Be Solved?

11.2.2.1 Key Factors

MVPs with Feedback Loop

The first step in value proposition is to propose a value proposition with a minimum viable product (MVP). The MVP is a version of the product that enables a full turn of the customer open innovation-based business model developing circle with a minimum amount of effort and the least amount of development time (Reis 2011, p. 77).

With this, the method most similar to the customer open innovation-based business model is the lean start-up. The build–measure–learn loop, as presented in Fig. 11.5, shows that the lean start-up develops the MVP created in the early stage into the completed form through continuous feedback.

The customer development model also has a pivot search loop that provides feedback with regard to customer discovery and customer validation and an execution loop that repeats the customer creation and company building processes (Blank 2013, p. 25). The customer development model is also similar to the customer open innovation-based business model developing circle presented here in consideration of the fact that the discovery of a customer describes a model gradually developed through a feedback loop.

11.2.2.2 Value Proposition for the Customer Open Innovation Business Model Circle

Value Proposition Template: Customer Open Innovation Business Model Circle

1. Describe your own problem that should be solved, and arrive at a concrete value proposition.

2. Describe your own requirement that should be realized, and arrive at a concrete value proposition.

3. Summarize the entire improvement by describing and giving shape to the concrete improvement in all aspects (quantitative, qualitative, functional, design, brand recognition, cost, cost reduction, and risk reduction improvement) and the improvement of the access to the value of the value proposition.

Briefly and clearly describe the concrete solution to a problem desired to be solved as a customer. Next, briefly and clearly describe the requirements additionally desired by a customer. This description should be based on concretely, objectively, recognizable, measurable, and realizable contents.

However, for the value proposition, it is necessary to make efforts clearly to suggest improvements in various aspects rather than only a single improvement and then to improve the points.

11.2.3 How Should the Technological System Be Constructed?

11.2.3.1 Key Factors

Key Resources, Key Activities, and Key Partnerships Are Included in a Technological System

It is meaningful that various key resources, such as human resources, material resources, and intellectual resources, that a customer business model developer has; key activities, such as problem solutions, production, and the platform operation; and key partnerships do not exist independently but are systematically combined.

Make Your Business Model Difficult to Copy

First, as presented in Table 11.2, it is important to make one's own business model complex in a technological system. It may not be impossible for a customer who

Table 11.2 A method to use to make a business model difficult to copy

Create a product or service that is patented, trademarked, or difficult to duplicate
Find a way to serve customers that is unprofitable or impracticable for competitors
Deliver products and services better, cheaper, or faster through a business process known exclusively by you or your company
Create an ecosystem where the ongoing use of your product is highly desirable or required

Source: Muehlhausen (2013, pp. 21–23) modified

Table 11.3 Three points for business model innovation

Connect: Business model innovation is a team sport
Catalyze something bigger than yourself
Enable random collisions of unusual suspects
Collaborative innovation is the mantra
Build purposeful networks
Together, we can design our future
Inspire: We will do what we are passionate about
Stories can change the world
Make systems-level thinking sexy
Transformation is itself a creative act
Passion rules—exceed your own expectations
Be inspiration accelerators
Transform: Incremental change isn't working
Tweaks won't do it
Experiment all the time
Off the whiteboard and into the real world
It's a user-centered world—designed for it
A decade is a terrible thing to waste

Source: Kaplan (2012, pp. 52–53) modified

intends to develop a customer open innovation-based business model to develop a better product, but it would be very difficult to do so. In addition, it is burdensome to construct an ecosystem in which the more a specific product or service is used, the higher the value of the use is from the perspective of a customer BM developer.

Thus, a customer business model developer should first consider securing the minimum legal defense for the business model through patent application or design registration processes, such as intellectual property for the technological system design. In addition, the development of a business model profitable for a customer business model developer and not profitable for other teams is required. However, this case also becomes a target of the patent application or design registration processes.

Technological System for Connecting, Inspiring, and Transforming

With regard to composing key activities, key resources, key partnerships, and a technological system, a customer business model developer needs to design a system that operates a business model based on a team rather than an individual, bringing about inspiration and accessing a new problem rather than simply improving the current situation, as presented in Table 11.3.

11.2.3.2 Technological System for the Customer Open Innovation Business Model Circle

Technological System Template: Customer Open Innovation Business Model Circle

1. Compose a technological system, including key resources, key activities, and key partnerships, while taking into consideration the creative combination between the technology and the market.

2. Form a technological system—that is, a creative combination between concrete and direct technology and the market—that cannot be easily copied, at least through patent application or design registration steps.

3. Compose a technological system in which a customer business model promotes connections between all members connected through the system, including the developer, and where the passionate participation of a customer business model developer and customers accelerate, inspiring significant change rather than gradual change.

11.2.4 Why Should You Establish a Start-Up?

11.2.4.1 Key Factors

Cash Flow of Start-ups

A customer business model developer should recognize that it takes much time to reach the breakeven period, as shown in Fig. 11.6.

Short-term Revenue and Long-term Distribution Cost

Thus, when designing the cost and revenue in a business model, cash flow should initially be considered thoroughly. In particular, for a customer business model, a creative cost and revenue system should be designed and presented to avoid financial risk during the business model development stage so as to allow the business model to become a sustainable start-up.

Among the various forms that constitute the revenue stream, such as product sales, usage fees, membership fees, rental fees, licensing, commission, and advertising, the selection criteria are identical to the overriding selection criteria of the price mechanism, such as fixed pricing and dynamic pricing (Osterwalder and Pigneur 2010, p. 37). This pertains to how rapidly revenue is generated.

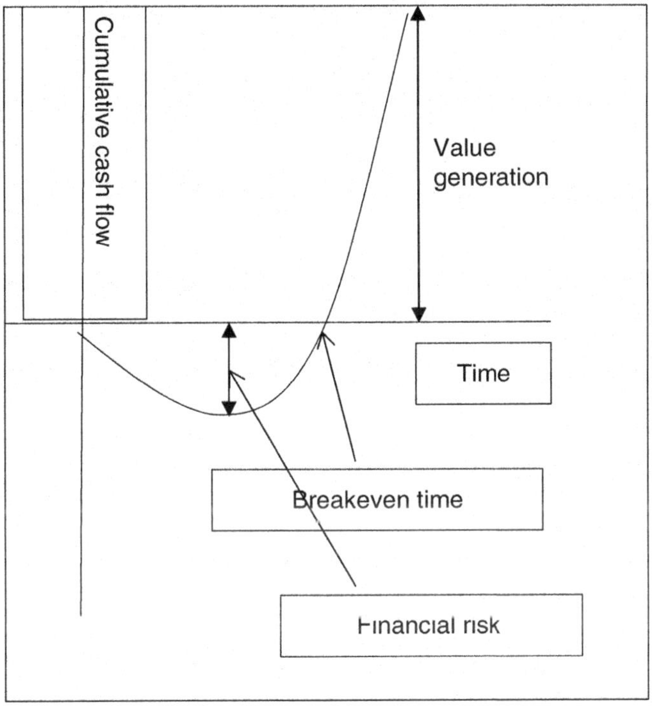

Fig. 11.6 Cumulative cash flow of start-ups (Source: Aulet and Murray (2013))

Meanwhile, there are diverse selections, such as a cost-reduction-led business model or differentiated value creation. For a customer business model, owing to the low capacity of the initial step, a cost design which may spread for a long time may be a prioritized matter of choice.

11.2.4.2 Cost and Revenue for the Customer Open Innovation Business Model Circle

Cost and Revenue Template: Customer Open Innovation Business Model Circle

1. A start-up based on a customer business model cannot create profits for a long time; in addition, it takes a long time before the breakeven period arrives. To avoid this, it is important to design costs and revenues creatively.

(continued)

2. Select a revenue form and price mechanism that rapidly generate revenue as soon as possible when selecting one from among various forms, such as product sales, usage fees, membership fees, rental fees, licensing, commissions, and advertisements, and/or when selecting the price mechanism, such as fixed pricing and dynamic pricing, during the design of a revenue stream.

3. Design a structure which prioritizes the fact that costs are needed for a long time when a cost structure is designed.

11.2.5 When and Where Does One Meet Customers?

11.2.5.1 Key Factors

The *Get, Keep, Grow* Funnel for Customer Business Models

If a customer becomes aware of, shows interest in, considers, and purchases an item, that is, if moving from "acquire" to "activate," as shown in Fig. 11.7, customers significantly decrease. However, as Get Customers is a significant customer business model, it is vital continuously to become a source of Get Customers through positive feedback with a viral loop. As a customer business model has an advantage with regard to willing nearby potential customers who share an interest identical to that of the developer, the interests of customers based on the owner's experience, in particular, during the Get Customers stage, should be secured, and concrete purchases should occur.

Through the *Keep Customers* process, which directly reflects the refinement of a business model developer's continuous interests and experiences, the requirements and expectations of customers should be continuously met with creativity. During this stage, a customer business model developer needs to group customers who have a critical mind similar to theirs and then must maintain the *Keep Customers* stage.

In addition, to operate a customer business model continuously, the *Grow Customers* strategy of upselling, in which the manner of meeting customers is based on high-end product lines, net-selling based on the product next to the existing product, and cross-selling based on a product contrary to the existing purchase, needs to be pursued. During this process, customers can be continuously secured,

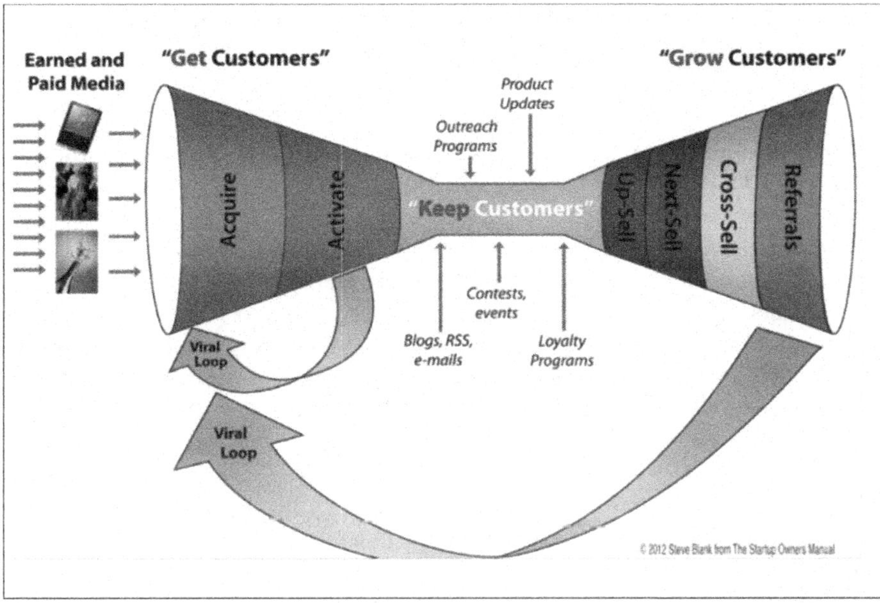

Fig. 11.7 The Get, Keep, Grow funnel in physical channels (Source: Blank and Dorf (2012, p. 126))

maintained, and expanded based on the process of the refinement of the initial critical mind as a customer and the application to other customers in various ways in a customer business model.

Sophisticated Customer Channels and Relationships Are the Highlights of Customer Business Models

Sophisticated designs of customer channels, such as direct management, the operation of a partner store, and the operation of online or mobile websites, represent the ultimate stage that allows the operation of a customer business model. Thus, whether a customer business model will succeed is determined by the weight of the attention given to this part.

Customer relationship designs in various forms, such as self-service, automation service, the community type, the co-creation type with a customer, customer community operations, standard support for each individual customer, and in-depth support of each individual customer, practically determine the sustainability of a customer business model. Thus, customer relationship design must start with the most preferred type of customer business model developer and must be concretely and directly operated (Osterwalder and Pigneur 2010, p. 35).

11.2.5.2 Channels and Customer Relations for the Customer Open Innovation Business Model Circle

Channels and Customer Relations Template: Customer Open Innovation Business Model Circle

1. Creatively maintain the process, starting with the inner view of a customer business model developer and obtaining a nearby potential customer. To do this, when organizing the channel of an off-line or online store or a mobile Web store, minimize the initial cost of the customer business model, and develop a concrete and direct method of meeting customers from experience, and then creatively apply the method.

2. In addition, the most proper method to share a critical mind between customers from a perspective of a customer is applied by a developer, and it is important to seek a means to allow customers to maintain their activities continuously. Find and apply the most valid and creative method, starting with the critical mind of a business model developer among various customer relations, such as the community, individual support, in-depth support, and customer co-creation processes.

3. During the customer business model development stage, concretely and directly design a method of expanding various customers to *Grow Customers* in advance, concretely and directly suggest this in the type of vision to be deployed in the near future, and form the momentum to maintain and continuously expand customers.

Research Question

1. Select any case from the newspaper and analyze through customer open innovation-based business model developing circle.
2. Develop your own business model through customer open innovation-based business model developing circle.

References

Aulet W, Murray F (2013) A tale of two entrepreneurs: understanding differences in the types of entrepreneurship in the economy. Available at SSRN 2259740.
Blank S (2013) The four steps to the epiphany. K&S Ranch, Pescadero
Blank S, Dorf B (2012) The startup owner's manual. K&S Ranch, Pescadero
Cho Y, Kim GY (2014) A case study on the innovative Cleaning Product, Haan Steam Cleaner. Admin Manag Res 7(2):51–73
Kaplan S (2012) The business model innovation factory: how to stay relevant when the world is changing. Wiley, Hoboken
Kodama F, Shibata T (2015) Demand articulation in the open-innovation paradigm. J Open Innov Technol Market Complex 1(1):1–21
Maurya A (2012) Running lean: iterate from plan A to a plan that works. O'Reilly Media, Sebastopol
Muehlhausen J (2013) Business models for dummies. Wiley, Hoboken
Osterwalder A, Pigneur Y (2010) Business model generation: a handbook for visionaries, game changers, and challengers. Wiley, Hoboken
Praceus S (2014) Consumer innovation at the base of the pyramid: emerging patterns of user innovation in a resource scarce setting. Springer Science & Business Media, Wiesbaden
Reis E (2011) The lean startup. Crown Business, New York

User Open Innovation-Based Business Model Developing Circle

12

Abstract
This chapter discusses cases and the development process and methods for customer open innovation-based business models (business models). Here, two cases of customer business models are introduced, followed by a detailed explanation of each of the five steps in the customer business model circle, with key factors and templates provided.

Keywords
Customer • Dyson vacuum cleaner • Han Kyung Hee steam cleaner • Customer business model circle

12.1 Cases

12.1.1 Nike+ (Nike Plus)

CEO Mark Parker of Nike stated, "Most runners were running with music already. We thought the real opportunity would come if we could combine music and data" (Ramaswamy and Gouillart 2010, p. 8). In line with this, Nike+ (pronounced *Nike Plus*), which was started in 2006, has formed a deep partnership with Apple in relation to its line of products, from the iPod to the iPhone 6s Plus, as well as the Apple Watch (Fig. 12.1).

First, with a run-tracking feature and the Apple Watch's Nike+ Fuel App, the Nike+ Running App, and the Nike+ Training App, a user can plot the distance, time, pace, and calories burned. It is also possible to issue running challenges to others through the Challenge Others feature. Second, running records and exercise experience can be shared with others through Facebook or Twitter. Third, through the

J.J. Yun, *Business Model Design Compass*, Management for Professionals,
DOI 10.1007/978-981-10-4128-0_12

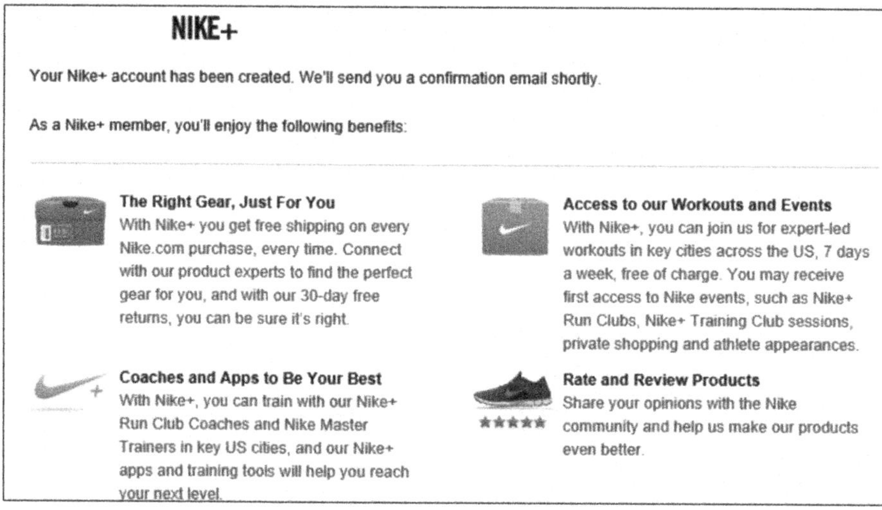

Fig. 12.1 Nike+ home page (Source: https://secure-nikeplus.nike.com/plus/)

Table 12.1 Nike+ as an excellent example of a co-creative engagement platform

Learn directly from the behavior of its customers
Generate new ideas rapidly
Experiment with new offerings quickly
Get direct input from customers on their running preferences
Build deeper relationships and trust with the community
Generate "sticker" brand collateral

Source: Ramaswamy and Gouillart (2010, pp. 11–12)

Nike+ Community, consumers are encouraged to present their ideas to improve existing products or even to suggest new products. The community systematically collects these ideas, comments, and opinions, which are reflected in Nike's product creation, eventually leading to the innovation of new products. Fourth, Nike+ systematically analyzes customers' data, where exercise and activity results are shared and made public. Nike+ makes use of this data actively to produce innovative products. Table 12.1 presents the analysis results of Nike+ as an excellent example of a co-creation case given that the ideas received from Nike customers through Nike+ become the source of the company's various innovations.

Nike+ is a platform that goes beyond serving as a community for users, as it also connects the direct/indirect requirements and expectations of customers to the production of innovative products and the innovation of existing products. In addition, this platform offers new means of entertainment and ideas to customers, such as the combination of running and music with competitive exercises and performances. In summary, Nike+ is a prime example of a company that provides new ways to attain customer satisfaction in product consumption, where the customers also have a say in the innovative efforts of the company.

As increasing numbers of people use smartphones and smart watches, users of Android phones as well as iPhones can download the Nike+ Fuel App, the Nike+ Running App, and the Nike Training App. The ecosystem of Nike+ is rapidly expanding. In addition, without the need to purchase a separate smart sensor, similar to the type built into Nike shoes, a user can simply download the Nike+ app and then use it as he or she runs while carrying a smartphone or wearing a smart watch. In this way, a user can enjoy Nike Plus, that is, potential customers, as well as existing customers of Nike, directly become objects of the company's innovation.

In the sports market, where the competition among many sports companies is already fierce, Nike uses Nike+ as a tool to introduce new products and services continuously and to demonstrate the company's consistent creation of new experiences for its customers. In the mature industry, a company actively presents new customer experiences, requirements, and expectations for the successful creation of new product and service business models (business models).

12.1.2 Starbucks

Two months after returning as CEO of Starbucks in January 2008, Howard Schultz launched the MyStarbucksIdea.com website, with these words:

> Welcome to MyStarbucksIdea.com. This is your invitation to help us transform the future of Starbucks with your ideas—and build upon our history of co-creating the Starbucks Experience together… So, pull up a comfortable chair and participate in My Starbucks Idea. We're here, we're engaged, and we're taking it seriously. (Ramaswamy and Gouillart 2010, p. 22)

Schultz organized the My Starbucks Idea platform for customers, as described in Fig. 12.2, with the motto "SHARE. VOTE. DISCUSS. SEE." As of January 2016, the site remains very active (Table 12.2).

The many ideas received from Starbucks customers through the platform pertain to products and services, falling into various categories, such as *Building a Community* and *Social Responsibility*. The business models of the latest product types implemented from customers' ideas are Bienvenue, the Almond Croissant (January 27, 2016), Simplifying a Favorite (New Green Tea Latte Recipe) (January 7, 2016), and Welcome Latte Macchiato to the Espresso Menu (January 5, 2016). In other services or processes prompted by customers' ideas, innovative business models such as Starbucks Music on Spotify (January 19, 2016; the music playlists played in Starbucks), the Digital Coffee Passport ((January 13, 2016) with this passport, customers gain access to stamps and other information and can keep track of each cup they drink, and more), and Starbucks Delivery in Seattle (December 2, 2015) are continuously created and applied.

Fig. 12.2 My Starbucks Idea (Source: MyStarbucksIdea.com)

Table 12.2 The reality of My Starbucks Idea (as of January 27, 2016)

Products idea	Coffee and espresso drinks	45,861
	Frappuccino beverages	6478
	Tea and other drinks	13,487
	Food	23,000
	Merchandise and music	11,283
	Starbucks card	23,434
	New technology	5942
	Other product ideas	14,562
Experience ideas	Ordering, payment, and pick-up	12,319
	Atmosphere and locations	23,108
	Other experience ideas	14,899
Involvement ideas	Building community	6774
	Social responsibility	11,421
	Other involvement ideas	6714
	Outside USA	2203

Source: MyStarbucksIdea.com

The cafe culture of Europe has a longer history than that of the USA. Moreover, it has been a long time since the cafe culture was combined with local culture to represent the principles of each European nation. This proves the difficulty of continuously creating new product and service business models in the coffee industry. However, through the My Starbucks Idea platform, Starbucks is able consistently to craft new products and services inspired by customers' ideas.

12.2 User Open Innovation-Based Business Model Developing Circle

As shown in Fig. 12.3, from the perspective of existing firms, in particular, small- and medium-sized enterprises (SMEs), it is the user open innovation-based business model developing circle wherein technology and the market are creatively recombined or additionally combined using the circle, which is composed of five business model elements according to the principle of the Five Ws and One H. No company can survive without new business model development, that is, the continuous promotion of creatively recombining technologies and the market. This demonstrates that companies, from small- and medium-sized business to some of the world's best known companies, such as Nokia and Kodak, can no longer survive without the continuous addition or renewal of business models. The most important factors in the innovator's dilemma, referring to moments when new technologies cause great firms to fail, are the failure of the open innovation of technology in the market and the new introduction of creative business models (Christensen et al. 2005).

All companies that exist in the world are considered as existing firms. Thus, the user open innovation-based business model developing circle is a methodology of

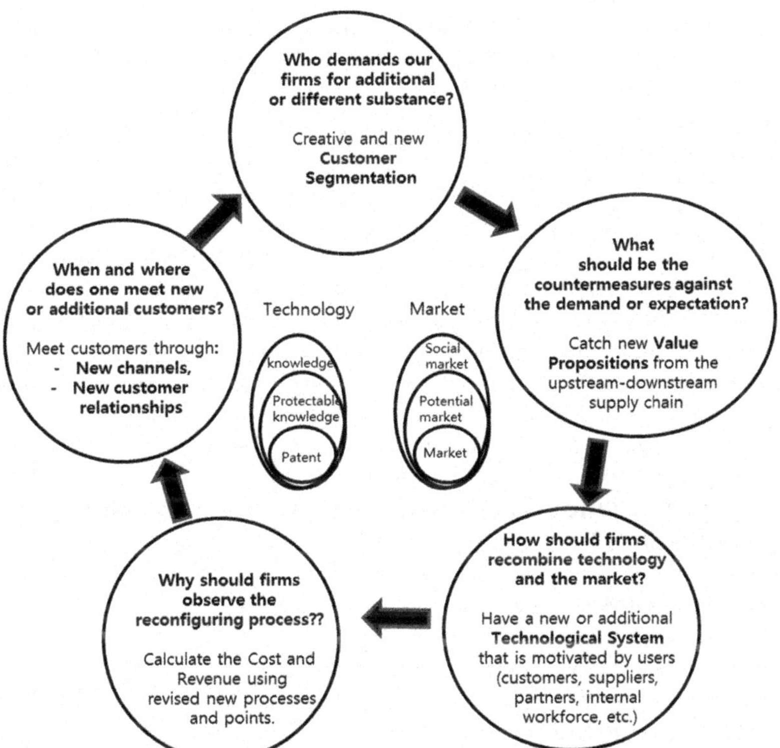

Fig. 12.3 (Existing firms') User open innovation-based business model developing circle

developing business models for all companies in the world—for all existing firms (Han and Cho 2015). With this, the main concern of start-ups which plan to introduce a new business model through the customer business model circle should be the methodology of developing an additional or transformed business model as an existing firm.

The characteristics of this methodology are the existing technology, the market, and the business model of the existing company. That is, there is a concrete and direct business that creates profits, grows, and maintains a company. The user business model circle depicted in Fig. 12.3 focuses on a creative approach that differs from, or acts as a substitute for, the existing approach, covering all five business model elements (customer segmentation, value propositions, technological system, revenue and costs, and customer encounters). For existing companies, the requirements and expectations from various points in the supply chain and those suggested by various sections in a company become the source of the creatively recombined business model of technology and the market regardless of the manufacturing or service field.

Thus, the user of the user open innovation-based business model circle as described in this chapter includes an entity that exists in the supply chain of a company, such as a customer, supplier, partner, retailer, and/or the internal workforce of a firm. The user of this chapter differs from the user of a user innovation; that is, the user serves as a subject of self-innovation created to realize one's needs as a customer of a product or service (Von Hippel 2005). First, a user of a user innovation is a user in the supply chain of an existing firm. However, the user of the user open innovation-based business model circle is the entity that innovates something for his/her use with a will for innovation among all customers of a specific product/service. Second, the former acquires users' comments and ideas through various activities, such as co-creation and collaboration, and an existing firm creatively recombines technologies and the market to create a new business model. However, the latter innovates something for the subsequent use of his/her assets based on ideas received from existing use experiences. The user of the user open innovation-based business model circle has a much wider activity scope than the user of a user innovation in terms of activity content, though the former is weaker in terms of activity.

12.2.1 Who Demands Our Firms for Additional or Different Substance?

12.2.1.1 Key Points

Customer-driven Supply Chain Open Innovation

Most importantly, the structure of the supply chain should be changed from a linear supply chain structure to a network of suppliers and customers. The effective alignment of corporate and supply strategies is essential if supply chains are to be

Table 12.3 The customer-driven supply chain open innovation process framework

Customer-driven guidelines			
Elimination of waste	Alignment of production with demand	Integration of suppliers	Creative involvement of the workforce
⬆	⬆	⬆	⬆
Customer-driven practices			
Process and value stream mapping Workplace organization Total productive maintenance (TPM)	Minimum travel and cellular layouts Pull systems Workstation balancing	Just-in-time deliveries Supplier relationship management and development activities Supplier rationalization	Team-based problem solving Quality circles Cross-functional training

Source: Lyons et al. (2012, p. 23) partially modified

designed to meet customers' needs (Lyons et al. 2012, pp. 3–4). A concrete method to realize this can be implemented through the customer-driven supply chain process shown in Table 12.3.

Finally, the core of the customer supply chain open innovation is to perform open innovation along the entire supply chain by carefully listening to the requirements of customers, the input of suppliers, and to various comments made by other firms in the supply chain.

Lead users or normal consumers have sufficient technology and knowledge in the field. They are constantly seeking new alternatives with high creativity with regard to product consumption. Moreover, they are, above all, innovators in the field, as shown in Fig. 12.4.

However, most of them directly engage in creative creation rather than merely continuing to provide feedback, as it is much easier creatively to combine technologies and the market, which exist in the world; such combinations can be created moreover at a low cost. Hence, focusing on the requirements from all users in the supply chain, i.e., customers, suppliers, the internal organization, the distribution system, and finally customer relations processes, is the most important starting point of the development of a user open innovation-based business model. A user, as a diverse customer in the supply chain, can also participate in proactive market research and R&D processes.

The Innovation Community

In addition, a user business model must make efforts to specify the requirements and expectations of the innovation community level shown in Fig. 12.4 and then to confirm the scope for customer segmentation. In the user open innovation-based business model circle, new customer segmentation initially leads to the development of a new business model which is qualitatively different from the

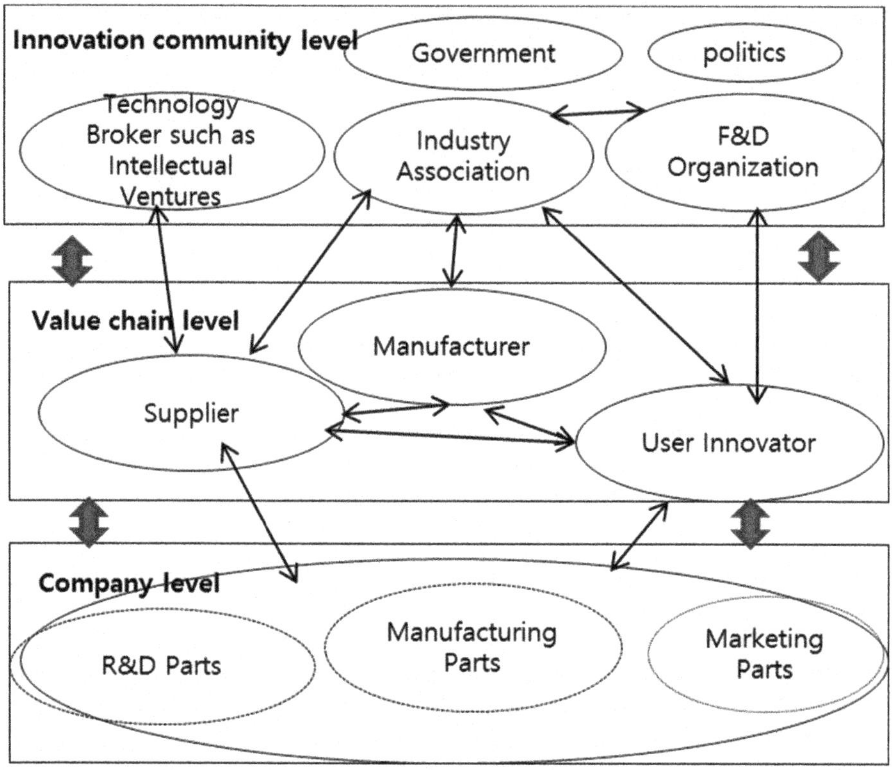

Fig. 12.4 Innovation community (Source: Fichter and Beucker (2012, p. 13))

existing one. Thus, the important source of differentiated customer segmentation can be secured from the requirements and expectations at the innovation community level.

12.2.1.2 Customer Segmentation for the User Open Innovation Business Model Circle

This stage, above all, specifies and concretely describes a customer group, differentiated from existing customers. In addition to the specification of a customer group in the value chain stage, innovation communities in three different dimensions are proposed, with customers accurately described as well.

Customer Segmentation Template: User Open Innovation Business Model Circle

1. Check and summarize a customer in the value chain.
   - Clarify the customer in the value chain. That is, summarize the requirements and expectations of a concrete subject, such as a supplier, a customer, and the internal workforce.
2. Check, distinguish, and summarize the existence of a customer in innovation communities beyond the value chain.
   - In addition, expand to the innovation community, confirm the level from the company level, the value chain level, or the innovation community level, and concretely describe a subject (i.e., a customer) of new requirements and expectations at that level.
3. Concretely define and express a new customer group that can be distinguished. Suggest a concrete customer group newly defined in a visibly understandable form.

12.2.2 What Should Be the Countermeasures Against Demands or Expectations?

12.2.2.1 Key Factors

Identify Other Customers by Identifying Nonconsumers and Overshot Customers

As shown in Table 12.4, from the various additional requirements and expectations of existing customers or the requirements of reality, nonconsumption areas without customers and overshot customer areas beyond the existing customer group are found. Afterward, these are concretely summarized to arrange countermeasures that can add to or substitute for the existing value proposition.

Gain Customer Insights

An integrated method among all methods used to approach a new or an additional value proposition is to find, accumulate, distinguish, and concretely aggregate new customer insights from the additional requirements and expectations received from various customers in the value chain. This is described in Table 12.5.

Table 12.4 Noncustomers and overshot customers

Calories	Definition	Contents
Nonconsumers	Nonconsumption as an absence of consumption refers to either people (nonconsumers) or contexts (nonconsuming contexts) where consumption is inhibited by certain barriers (Anthony 2008, p. 46)	There are four constraints associated with noncustomers, as follows: Skill-related constraints Wealth-related constraints Access-related constraints Time-related constraints
Overshot customers	"An overshot customer is a particular customer segment for which existing products or services are more than good enough. Overshooting occurs when a product or service has performance that a customer doesn't need, and therefore doesn't value" (Anthony 2008, p. 66)	Three specific approaches that can help identify signs of overshooting: Direct interactions with customers Margin, price, and share analyses Analyses of recent product introductions

Source: Anthony (2008, pp. 45–86). Main contents are summarized

Table 12.5 Techniques to gain customer insights

Techniques	Contents	Facts to be noted
Data detective	Build on existing work with research Secondary research reports and customer data Data outside existing industry	Data is data Do not confuse it with the real world
Journalist	Talk to (potential) customers to gain customer insights	Customers may say *A* in an interview but do *B* in the real world
Anthropologist	Observe customers in the real world to gain useful insight from them	It takes time and money It is difficult to gain customer insights related to new ideas
The impersonator	Be your customer, and actively use your products and services	Sometimes, it is impossible to become a representative of an actual customer or is not even possible to attempt to do so

Source: Osterwalder et al. (2015, pp. 106–107)

Accurately capturing and systematically accumulating the items clearly or implicitly suggested or required by a customer from the perspective of a data detective, a journalist, an anthropologist, and an impersonator can become a new value proposition.

12.2.2.2 Value Proposition for the User Open Innovation Business Model Circle

Value Proposition Template: User Open Innovation Business Model Circle

1. Find and concretely summarize the nonconsumption and overshooting of a customer among existing customers to suggest a concrete and new value proposition.
   - In the additional requirements and expectations of existing customers, concretely describe their new requirements and expectations or problem–solution wish areas with the starting point of the four constraints of wealth, access, time, and skill.
   - List, summarize, group, and systematically aggregate concrete requirements and expectations in a different dimension, overshooting beyond the current customer level.

2. Concretely identify various customers' requirements in the value chain in depth, and propose a new value proposition through co-accumulation and aggregation.
   - Describe the views of a data detective, a journalist, an anthropologist, and a creative impersonator, and fully identify, analyze, aggregate, and show the multifaceted section of value for additional or new requirements of customers.

12.2.3 How Should Firms Recombine Between Technologies and the Market?

12.2.3.1 Key Factors

Co-creation Platforms

> Co-creation is the joint creation and evolution of value with stake-holding individuals, intensified and enacted through platforms of engagements, virtualized and emergent from ecosystems of capabilities, and actualized and embodied in domains of experiences, expanding wealth–welfare–well-being. (Ramaswamy and Ozcan 2014, p. 14)

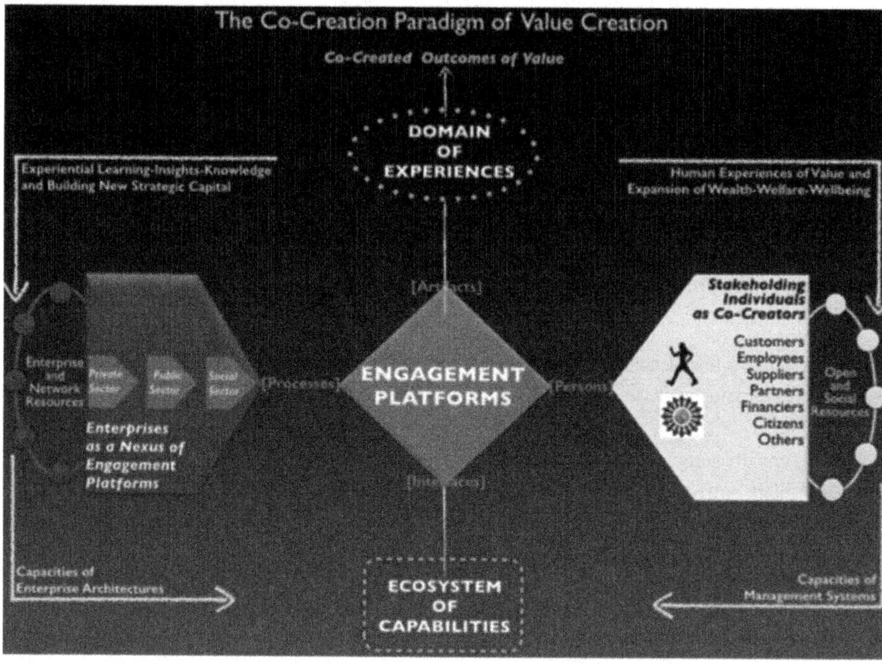

Fig. 12.5 The co-creation paradigm of value creation (Source: Ramaswamy and Ozcan (2014, p. 29))

As shown in Fig. 12.5, the co-creation platform enables the ecosystem of capabilities, an engagement platform, and the domain of experience to be activated based on the participation of stakeholders, leading to the co-creation of outcomes of value. The co-creation platform can continuously increase the value creation opportunities by expanding the space of experiences, the scope and scale of interactions, stakeholder relationships in the ecosystem, and linkages among engagement platforms.

Co-creation Sectors and Firms

As shown in Table 12.6, co-creation refers to the characteristics of open innovation and business model development, generally emerging in most industries, including primary, secondary, and tertiary industries and mature industries of both service and manufacturing.

As presented in Table 12.5, from global companies such as Apple or Google to start-ups, companies of various sizes and in various fields and regions use the concept of co-creation business model development. For business model development based on user open innovation from various customers in the value chain, existing companies can receive motivation to take creative approaches with regard to system formation from those cases in the world of new technological systems. If a user open innovation business model in a different sector is referred to during the introduction of a technological system in the business model development area to be added or transformed, one may gain a very useful result.

Table 12.6 Examples of co-creation sectors and firms

About	Contents
Sectors	Agricultural life sciences, automotive, capital intensive equipment Commodities, consumer durables, electronics, energy, entertainment, fashion Fast-moving consumer goods (FMCG), financial services Health care, industrial goods and services, IT services, manufacturing and contract services, IT services, manufacturing and contract services, media, pharmaceuticals Professional services, public and citizen sectors, retail, social sector, software Telecom, travel
Firms	Abb, Amazon, Apple, Ashoka, BEME (Chile), Brother, Caja Navarra (Spain), Camiseteria (Brazil), Cisco, Blub Tourism (Japan), Credit Agricole (France), Crushpad, Dassault Systèmes, Dell, ERM GE Healthcare, GlaxoSmithKline, Google, HCL Technologies (India), Hindustan Unilever (India), IBusiness Model, Infosys (India), InnoCentive, Intuit, ITC (India), Jabil Circuit, Kaiser Chemicals (Disguised), La Poste (France), LEGO, Mozilla, Nestle, Nike, OASIS (S. Korea), Orange (France), Rio Grande do Sul (Brazil), SAP, SEBI, Shell, Sony, Starbucks, TiVo, Toyota Scion, Wacoal (Japan), ZARA (Spain), WordPress, Local Motors, Kindle Direct Publishing, Lezhin Comics (S. Korea), Socar (S. Korea), Fooducate (S. Korea), Scoutzie, Kukka (S. Korea), Songza, Datacoup, Threadless, ShareBling, BurudaConcert (S. Korea), 4Food, Memebox, Ryanair, New York Times Digital, eBay, Wikipedia,

Source: Ramaswamy and Gouillart (2010, p. 31) partially modified; in the Firms section, 18 items are added through an analysis by the author in Nam et al. (2016); two items are added through an analysis by the author

12.2.3.2 Technological System for the User Open Innovation Business Model Circle

Technological System Template: User Open Innovation Business Model Circle

1. Implement an engagement platform in a system for co-creation.
   - To do this, construct the capability ecosystems based on generativity, likability, evaluability, and inclusivity.
   - In addition, build experience domains based on access, dialog, reflexivity, and transparency.
   - Finally, construct the engagement platforms with intentionality, transformability, integrativity, and creativity.

2) Through analyses of various cases of industries in the value chain during the process of creating a system for new value creation, form an additional or a new system.
   - Amazon, Google, eBay, Procter & Gamble, Threadless, Boeing, Brewtopia, Crowspirit, Wikipedia, YouTube, Twitter, Apple, Pandora Media, Netflix, and others are examples.

(continued)

3) Independently materialize the technological system for the creative recombination of technology and the market based on the requirements and expectations of users.
   - In particular, in the supply chain, find the closest customer in the current stage for a new business model, and add the additional creativity and novelty of a technological system through his/her active participation or cooperation.

12.2.4 Why Should Firms Observe the Reconfiguring Process?

12.2.4.1 Key Factors

From a Product-driven Business Model to a Service-driven One

As shown in Fig. 12.6, as various customers in the supply chain or value chain take part in co-creation, the business model based on the existing manufacturing industry is transformed into a service-added business model. For a manufacturing-centered business model, there is little room for co-creation in the value chain, and margin arises during the final stage of the value chain. However, in the user open innovation business model, value is additionally created throughout the value chain of customers, and each value leads to margin. At the bottom of Fig. 12.6, the darker part is the creation of additional value and margin. For example, for the transformation to the services approach in a food business operated by a chef in the product-focused grocery business, in a grocery store, all profits and margins are created from the food products, where value is added and integrated, and the cost is the sum of each contribution in the value chain. For a restaurant operated by a chef, all value is created from the sum of all services provided by the restaurant. Food products are also the result of service additions. In addition, the cost stems from the entire process of value addition rather than from the existing value chain.

Open Services Innovation Concept Map

As in Fig. 12.7, the entire process of a new business model can be redesigned based on a service using the open service innovation concept map. It is also possible systematically to redesign the cost and revenue characteristics during this process.

That is, systematic access to a business model is materialized through service setting, cocreation, and the open innovation of an additional business model. Eventually, the transformation or addition of a new business model can be realized.

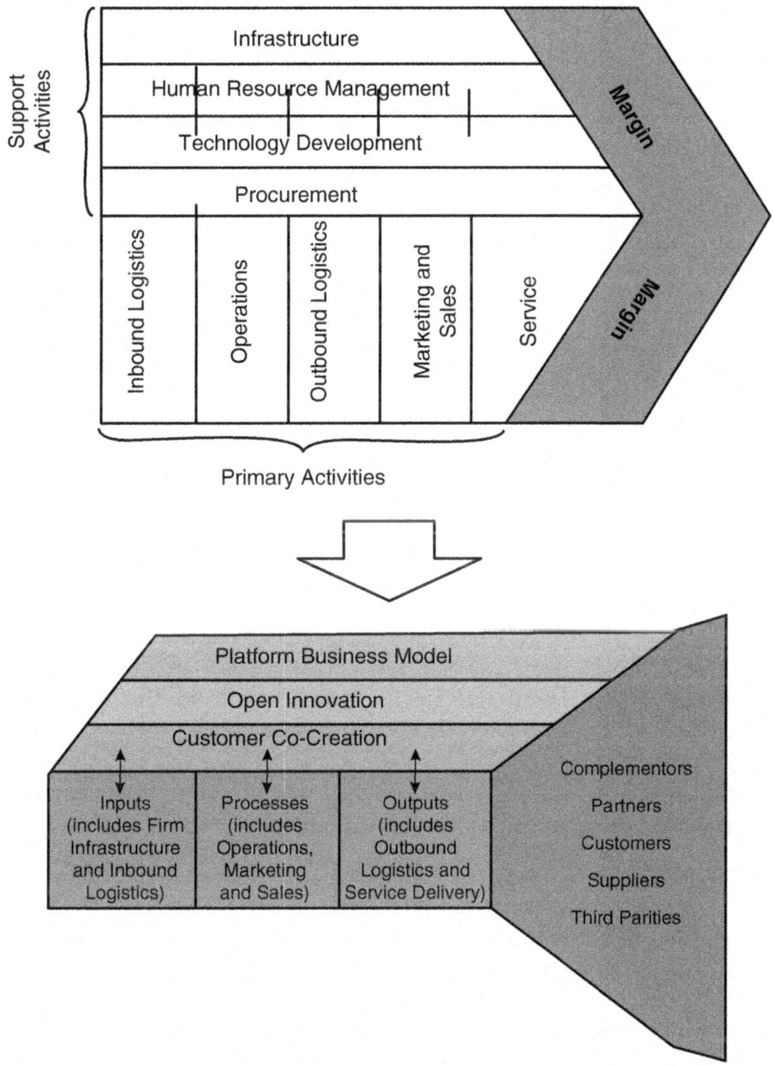

Fig. 12.6 From product to service in business models (Source: Chesbrough (2011, pp. 33, 35); creatively combined by the author; the upper part is a reprinted part of Porter (2008) in Chesbrough (2011))

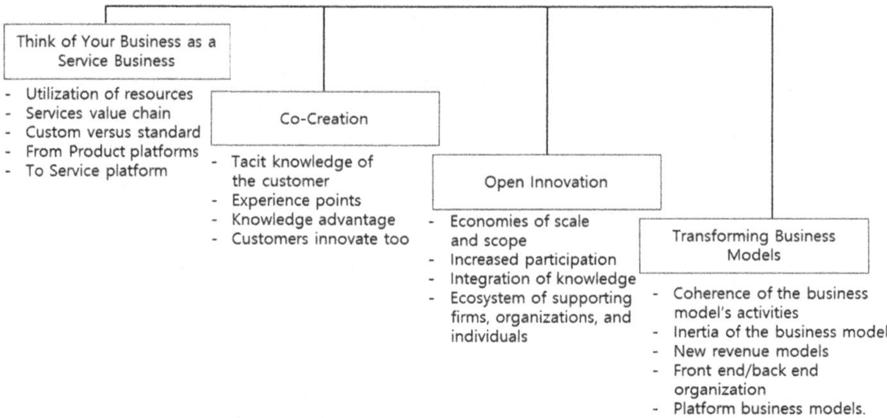

Fig. 12.7 Open service innovation concept map (Source: Chesbrough (2011, p. 110); the figure is partially modified and redefined)

12.2.4.2 Cost and Revenue for the User Open Innovation Business Model Circle

Cost and Revenue Template: User Open Innovation Business Model Circle

1) Drastically move from product to service when a cost is designed. Move from service to an advanced service + product.

   - For the cost and revenue of a new business model, a system is designed by qualitatively changing a product to develop a business model in a qualitatively different dimension and creating performance in a quantitatively different stage.
   - The user open innovation-based business model type indicates the servitization of a product.
   - In this case, the cost is generated during the co-creation process by users and by the value chain. In addition, revenue is generated during the previous stage of user co-creation rather than being created as margin during the final stage of the value chain.
   - Drastically design the cost and service to meet the service open innovation.

(continued)

2) Escape from existing customers and revenue. If adhering to the existing profit system, a new business model cannot be created, and escape from the existing business model is not possible.
   - Introduce the open service innovation concept map, drastically concretize a new business model, and show the corresponding cost and revenue.

3) Finally, set a quantitative goal and redesign the cost and revenue to achieve the aim.
   - This represents the key to the user value proposition redesign process.

12.2.5 When and Where Does One Meet New or Additional Customers?

12.2.5.1 Key Factors

Open Source as a Revitalizer of Open Innovation and Creative Business Models

As shown in Fig. 12.8, open source activates open innovation, which is an open connection between technology and the market, and the business model, which is a creative combination of technology and the market. Open source as an inspired manner of collaboration among motivated individuals working in "communities" is a voluntary community of individuals in various fields, starting from SW development; it plays a role in the potential and current customers of companies in the value chain.

It is reasonable that meeting customers (i.e., users) in the value chain or supply chain is differentiated and individualized depending on the customer, as shown in Fig. 12.9. However, for the basic work of the entire process in the value chain, a standardized process is essential. Thus, when organizing the customer channel and customer relations, it is crucial to balance between customer-customized differentiation and standardization.

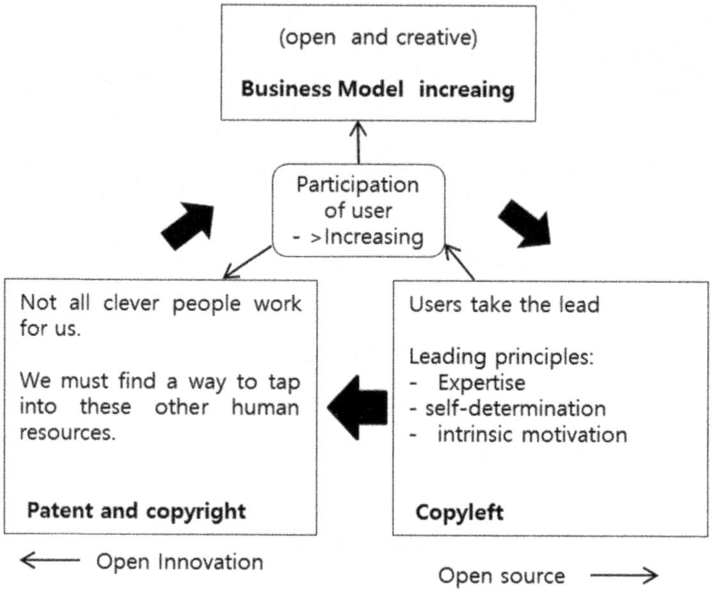

Fig. 12.8 Open source revitalizes classical open innovation and creative business models (Source: Bloem et al. (2007; 27) modified)

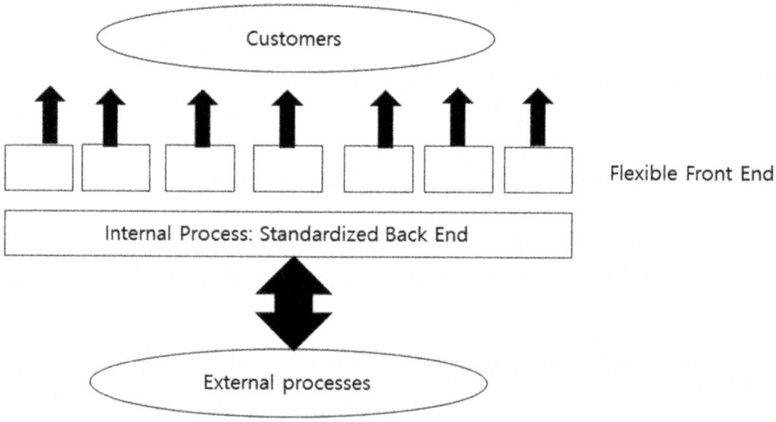

Fig. 12.9 Customized front end and standardized back end (Source: Chesbrough (2010, p. 21))

12.2.5.2 Channels and Customer Relations for User Open Innovation Business Model Circle

Channels and Customer Relation Template: User Open Innovation Business Model Circle

1. Concretely and independently define and confirm the channel and customer relations where a potential customer encounters a product in the current supply chain.
   - With open-source access, by activating the participation of all users in the supply chain, the opportunity to realize creative customer channels and customer relations can be secured.

2. Make the best use of the existing channels and customer relations to apply them to a new business model creatively.
   - The criterion is the balance between differentiation and standardization. Balance the construction of differentiation between users and the customized channel and customer relations and that of the standardized channel and customer relations created according to a standardized procedure set to a certain standard.

3. Among various channels and customer relations considered for open sources, consider the differentiated section according to the situation and the standardized case that takes into consideration the characteristics of the value chain; however, clarify the differentiation and identity between the current channels and customer relations to secure the creativity of a new business model.

Research Question

1. Select any firm at newspaper, and analyze the business model of the firm through user open innovation-based business model developing circle.
2. Build up your own business model through user open innovation-based business model developing circle.

References

Anthony SD (2008) The innovator's guide to growth: putting disruptive innovation to work. Harvard Business Press, Boston

Bloem J., van Doorn M., van Ommeren E. (2007). Open for business: open source inspired innovation. Bariet, SOGETI

Chesbrough H (2010) Open services innovation: rethinking your business to grow and compete in a new era. Wiley, Hoboken

Chesbrough H (2011) Open services innovation. Res Technol Manag 54(6):12–17

Christensen JF, Olesen MH, Kjær JS (2005) The industrial dynamics of open innovation—evidence from the transformation of consumer electronics. Res Policy 34(10):1533–1549

Fichter K, Beucker S (2012) Innovation communities: teamworking of key persons-a success factor in radical innovation. Springer Science & Business Media, Berlin

Han J, Cho O (2015) Platform business eco-model evolution: case study on KakaoTalk in Korea. J Open Innov Technol Market Complex 1(1):1–14

Lyons AC, Mondragon AEC, Piller F, Poler R (2012) Customer-driven supply chains: from glass pipelines to open innovation networks. Springer Science & Business Media, Berlin

Nam DI, Kim JH, Ahn HJ, Jeung JH, Lee KY (2016) Business Model Story 101 Successes. Hans Media. (Written In Korean)

Osterwalder A, Pigneur Y, Bernarda G, Smith A (2015) Value proposition design: how to create products and services customers want. Wiley, Berlin

Porter ME (2008) Competitive advantage: creating and sustaining superior performance. Simon and Schuster, New York

Ramaswamy V, Gouillart FJ (2010) The power of co-creation: build it with them to boost growth, productivity, and profits. Simon and Schuster, New York

Ramaswamy V, Ozcan K (2014) The co-creation paradigm. Stanford University Press, Stanford

Von Hippel EA (2005) Democratizing innovation. MIT Press, Massachusetts

Engineer Open Innovation-Based Business Model Developing Circle

13

Abstract

This chapter explores cases which exemplify the development process and methods of engineer open innovation (OI)-based business models (BMs). To provide an introduction to engineer open innovation business models, three cases, i.e., that of Elon Musk with Tesla Motors and SpaceX, that of Larry Page and Sergey Brin with Google, and that of Steve Jobs and Steve Wozniak with Apple, are presented.

In addition, a detailed explanation of each of the five steps in the engineer business model circle is provided with key factors and templates.

Keywords

Engineer • Elon Musk • Larry Page • Sergey Brin • Steve Jobs • Steve Wozniak • Engineer business model circle

13.1 Cases

13.1.1 Elon Musk with Tesla Motors and SpaceX

First, Elon Musk was confident based on his engineering knowledge that the technical base for the electric vehicle and private spacecraft markets is already mature (Table 13.1).

With this confidence, he made a unique attempt to access the markets. In the electric vehicle market, he decided initially to manufacture high-end electric sports cars, in contrast with the actions of existing automobile companies that produce automobiles with low to high specifications. It was his belief that the high cost, which is considered as a major hurdle for electric vehicles, could fully balance the

J.J. Yun, *Business Model Design Compass*, Management for Professionals,
DOI 10.1007/978-981-10-4128-0_13

Table 13.1 Tesla Motors and SpaceX

Category	Tesla Motors	SpaceX
Critical mind of engineers	Technical factors involved in the manufacturing of electric vehicles, such as lithium-ion batteries, are further developed	Among the technologies already introduced, it is possible to create a private spacecraft market by securing the top relevant technology, manufacturing simple aircraft units, and using rocket propellants
Unique access to the market	Tesla Motors adopts the strategy of accessing the vehicle market, different from those of existing vehicle companies such as those who make top electric sports cars, roadsters, advanced electric sedans, X vehicles, universal electric vehicles, Model 3 types, and others As each stage develops, this company expands mass production and economies of scale based on its accumulated technology	SpaceX intends to create a private spacecraft market with a cost of one tenth of the cost of launching existing space shuttles SpaceX improves the momentum of rockets by unifying rocket engines as a "Merlin engine" to decrease parts and manufacture costs significantly. They also increase the number of engines built
Short-term technology accumulation through open innovation	Tesla Motors introduced the technologies of electric motors and conversion devices from AC Propulsion and design-related technologies from Lotus Cars It established the principle: "The important thing is we do not outsource matters but directly handle them inside the company as much as we can" Advanced technologies such as battery packs connecting 6831 lithium-ion batteries and smart electric vehicle charging systems were accumulated over a short period of time	SpaceX differentiated its simplified design and organization structure from the existing NASA access type SpaceX introduced the phenolic-impregnated carbon ablator (PICA), which is a spacecraft insulation from NASA, and developed PICA-X, its own insulation only for SpaceX's spacecraft Dragon, to secure the technical basis for reusing spacecraft units SpaceX also adopted and applied the "friction stir welding" technology developed by The Welding Institute (TWI), a British research institute, to the process of manufacturing the fuel tank of the Falcon rocket, leading to a significant reduction of the possibility of defects

Source: Morris (2014), 竹内一正 (2013) summarized

advantages provided by top-class vehicles, such as first-rate functions, strong output and design, and social value as an eco-friendly cars. He created a new high-end electric vehicle market. In addition, with various electric vehicle system technologies accumulated in this market, he applied them to the top-class sedans "Tesla

Model S" and "Tesla Model X." This allowed him significantly to decrease costs and secure profits through mass production. Currently, he is guiding the universal electric vehicle called "Tesla Model 3" through the final production stages.

With regard to technical innovation, he attempted to acquire external cutting-edge technologies around the world through open innovation and made efforts internally to accumulate important technologies, such as Tesla's battery pack and SpaceX's PICA-X, to secure various system technologies over a short period and greatly improve the competitiveness of the two companies. In addition, with the strategy of continuously surviving the competition in the form of large automobile giants, NASA, or aerospace companies such as Boeing, he expands the business models by creating profits along various points of the value chain in addition to selling the finished products. For example, for TESLA, there are its battery pack, charging module sales, and finished vehicle sales, while in the case of SpaceX, there are its rocket launch service sales, spacecraft launch service, and rocket service. That is, to improve the technical base, he simplified technologies, such as lithium-ion batteries and Merlin Rockets, to accumulate technologies over a short period of time and pursued diversification along various points of the value chain for profit creation.

13.1.2 Larry Page and Sergey Brin with Google

In 1996, when Sergey Brin and Larry Page visited the Office of Technology Licensing of Stanford University, there was fierce competition in the field, where search engines such as Yahoo, AltaVista, and MSN were developed as search portals. That is, when Brin and Page developed a new search technology, several search engine giants already dominated the market. However, based on the "page rank" technology shown in Fig. 13.1, the two young men introduced the Internet search method that returned Web pages based on the closest relationship to a search term and arranged the search results based on this status. They considered that the "page rank" idea would be fully competitive in the market, where existing companies focused on search engines with low accuracy based on a specific word (Brin and Page 1998). In the already mature search engine market, they created a new market based on technology innovations and recreated the search engine market into a blue ocean market during its growth period (Jarvis 2011, p. 24).

In addition, they built an advertisement system with a new concept called *keyword search advertisements*, which posted advertisements on sponsor links in the upper right of a search result page in the order following the highest to the lowest bidder. However, Google added "popularity" on top of the cost as a factor for determining the ranking of the advertisement arrangement. Although a "sponsor link" was placed at the bottom of the search results page, if it was clicked by many Internet users, its arrangement ranking could increase despite the fact that its advertisement cost was lower relative to those of the others (Schmidt and Rosenberg 2014, p. 69). Google led the search engine advertisement market, which was conventionally unreasonable to the "technology" world (Auletta 2010, p. 143), which

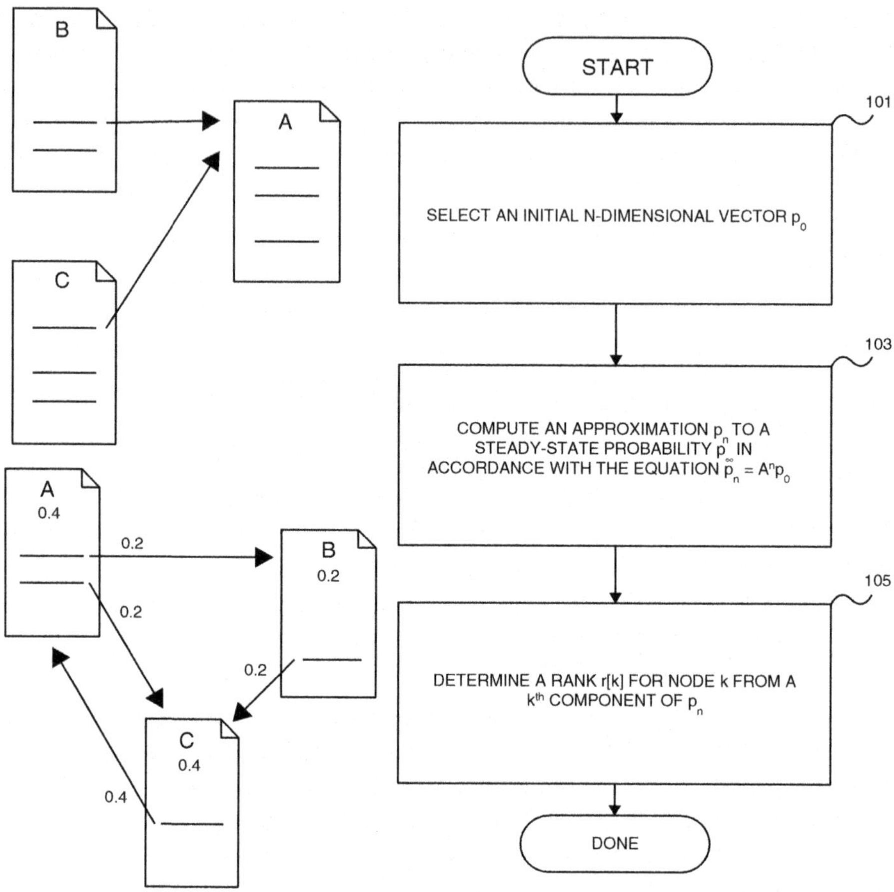

Fig. 13.1 Page rank concept picture (Source: US (Page 2001) patent US6285999 B1, method for node ranking in a linked database)

was then led by GoTo.com. As this company was bought by Yahoo, GoTo.com handed over its 2.70 million stocks, worth USD 230 million, to Yahoo, and the patent dispute over "keyword advertisements" was ended. The combination of the new search engine technology with the new search advertisement technology also created a new search advertising market.

This act of entrepreneurship allowed Google's further growth as a leader of the second IT innovation through market creation with continuous technology innovations based on strategies that welcome open innovation such as "bets on technical insights, not market research," "a period of combinatorial innovation," and "default to open, not closed" (Schmidt and Rosenberg 2014, pp. 67–94). Large search engine firms, such as Google, should carry out endless technology innovations to create additional advertising revenues and new businesses. For example, Google did not stop with its search engine but has continued to explore other innovative services.

Examples include Google News, which classifies stories according to their subject rather than source; Chrome, which is Google's open-source browser redesigned according to the browser speed; Knowledge Graph, which systematically structures Internet data; Google Translate, which improves the quality of translations without the help from the user; and Hangouts, which allow connections and video conferences regardless of how devices and videos are connected.

13.2 Engineer Open Innovation-Based Business Model Developing Circle

As shown in Fig. 13.2, from the perspective of engineers, circulating the cycle according to the principle of the five Ws and one H and creatively combining technology and the market can pave the way for the completion of a new business model. Before its proposal, this model began as a result of engineers' critical thinking based on their technical expertise. With regard to the characteristics of this business model, engineers criticize the current market, technology, and a business model based on

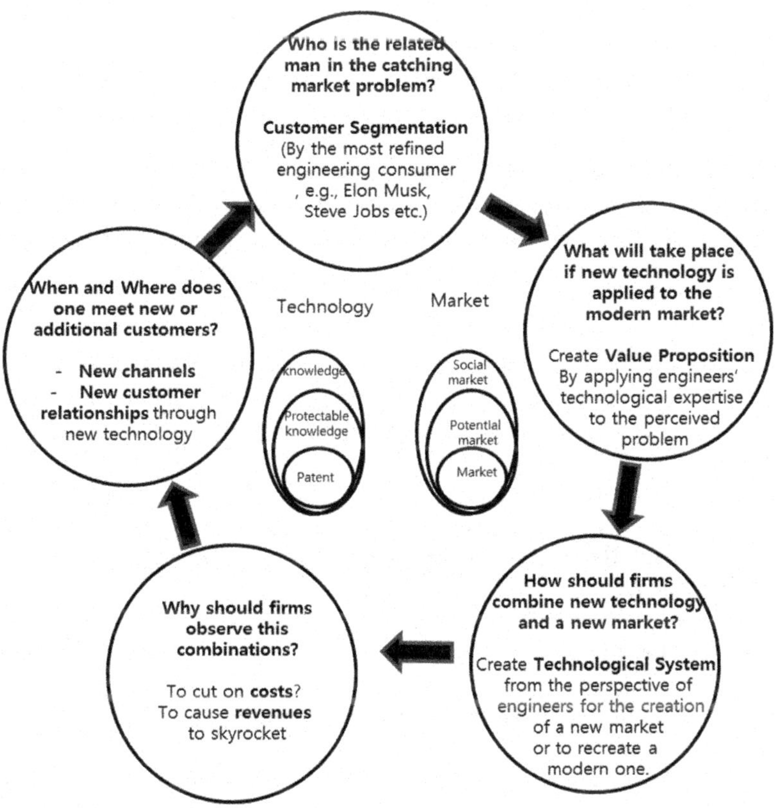

Fig. 13.2 Engineer open innovation-based MB developing circles

Table 13.2 Qualitative relationship between technology and business models

Contents	Logical background
Inventions differ from innovations	It is not an innovation until it delivers real value to a consumer
Ideas, inventions, and new technologies are the lifeblood for innovation	We must continue to invest in basic and discovery research, but only doing this is not sufficient
We must improve our ability to move inventions out of the lab and into the real world	We can solve problems and deliver value in the real world
Business model innovation is the key to realizing the full potential of new technology	R&D for new business models is imperative to remain competitive, harness technology, and deliver more value with fewer resources

Source: Kaplan (2012, pp. 111, 112)

their technical expertise, and this critical mind allows the development of a new business model.

Contrary to normal customers, engineers, including foreign R&D centers, are developers of new business models (Patra and Krishna 2015). They conceptualize the open innovation and, ultimately, a new business model in terms of new technology in the research stage, which is the starting point of the knowledge funnel. This is in contrast to the customer open innovation business model, which considers the open innovation and new business models from a market perspective that is differentiated from the existing situation composed of new consumers in the market stage, which is the end of the knowledge funnel.

In addition, engineers themselves serve as unique and sophisticated customers, forming a significant part of a certain market. They can identify the problems of the existing market, propose new customer segmentations, and explore new value propositions from the point of view of a sophisticated customer with technical expertise.

That is, the engineer open innovation business model circle is composed of five elements for the development of a creative business model based on new open innovation from the perspective of a sophisticated consumer with excellent insight in the relevant technology.

13.2.1 Who Are the Relevant Persons When Chasing the Market Problem?

13.2.1.1 Key Points

The Relationship Between an Invention or Technology and Business Models

As shown in Table 13.2, an invention is not innovation. If an invention has practical value to consumers, it becomes an innovation. Here, it is clear that a new technology is a key element of innovation. Thus, it is necessary to invest fully in basic and

Table 13.3 Four aspects of obstacles that prevent efficient transactions of technology between firms

Internal obstacles such as the NIH syndrome
Firms' innovative cultures
Obstacles related to interfirm relations
Obstacles related to the institutional structures of the MfTI in which the transactions take place

Source: Tietze (2012, p. 9)

discovery research. Only when a technology is implemented from a laboratory to the real world can it handle the problems of consumers and provide new values. That is, a value proposition is possible through technology. Finally, it is business model innovation that infinitely realizes the potential of new technology. However, a creative business model remains an important target of investment in R&D. It is necessary to expend all effort to invest in R&D for business models to provide more value with fewer resources to consumers, to create more revenue, and fully to use technologies.

Thus, on the assumption that engineers clearly understand the differences between technology and the market, a critical mind with regard to the existing market is an important starting point for an engineer open innovation business model.

There are many obstacles that firms face during transactions in the market for technology and innovation (MfTI), such as "the not-invented-here (NIH) syndrome, valuation difficulties when assessing market prices, and identifying buyers" (Tietze 2012, p. 9).

Engineers should ponder the four aspects of obstacles presented in Table 13.3 when comprehending problems or carrying out new customer segmentation activities.

13.2.1.2 Customer Segmentation for the Engineer Open Innovation Business Model Circle

Customer Segmentation Template: Engineer Open Innovation Business Model Circle

1. Comprehend the market problem from the perspective of an engineer.
   - If focusing on the qualitative relationship between technology and the market, an engineer can detect more creative market problems.
   - Based on a clear awareness of the difference between technology and business models, describe the market problem as conceptualized by an engineer.

(continued)

2. Describe the expected consumer when a problem is solved.
   - Considering the four aspects of obstacles that prevent transactions, think about the consumer and the solution to the problem. Although a problem can be handled technically, it is often difficult to secure the technology. However, as an engineer, it is possible to do both.

3. Describe customers in a concise manner.

Table 13.4 Open innovation-intensive areas

Issue	Contents and reasons
Autonomous innovation > systemic innovation	Autonomous innovation—ICT sector, pharmaceuticals Systemic innovations—chemicals, steel, railroads, petroleum
Automotive and aerospace	First- and second-tier suppliers play a growing part in the innovation process Manufacturers in these industries (often MNES) have shifted many innovative activities to their supplier companies over the years
Open innovation-oriented industries' structural characteristics	High globalization High-tech intensity High-tech fusion High speed of the appearance of new business models High knowledge leveraging effect for a competitive advantage of companies
Technological regimes	Opportunities—faster and more pervasive technological change Appropriability conditions High cumulativeness Knowledge base: multidisciplinary and cross-functional complexity

Source: Acha (2008), Chesbrough and Teece (2002), Gassmann (2006), Herstad (2007)

13.2.2 What Will Take Place If New Technology Is Applied to the Modern Market?

13.2.2.1 Key Factors

Open Innovation-friendly Industries and Technologies

When engineers apply their technical advancements to an existing market, it becomes easy to develop a new practical value proposition by mainly accessing open innovation-intensive areas, as shown in Table 13.4. These include industries

Table 13.5 15 Business model innovation principles

Connect: Business model innovation is a team sport
1. Catalyze something bigger than yourself
2. Enable random collisions of unusual suspects
3. Collaborative innovation is the mantra
4. Build purposeful networks
5. Together, we can design our future
Inspire: We do what we are passionate about
6. Stories can change the world
7. Make systems-level thinking sexy
8. Transformation is itself a creative act
9. Passion rules—exceed your own expectations
10. Be inspiration accelerators
Transform: Incremental change isn't working
11. Tweaks won't do it
12. Experiment all the time
13. Get off the whiteboard and into the real world
14. It's a user-centered world—design for it
15. A decade is a terrible thing to waste

Source: Kaplan (2012, pp. 32–33)

with easy autonomous innovations rather than system innovations, such as the automotive or aerospace industries, sectors with strong structural characteristics of strong open innovation, and industries dominated by a technological regime with strong open innovation. In such areas, it is easy to form open connections between technology and the market. Thus, it is very possible creatively to combine technology and the market, that is, to develop a creative business model.

Keep the Business Model Innovation Principles in Mind

When engineers apply a technology to an existing market, they initially must keep in mind the business model innovation principles given in Table 13.5. With teams rather than individuals, based on the factors that can create personal passion while also motivate reasonable types, and by adopting a radical approach rather than an incremental one, new value propositions can be proposed for markets and consumers, which may be ignored by engineers.

Build Adjacent Innovation Platforms

In this stage, while engineers maintain the current business model and construct an adjacent innovation platform, they find new methods that can create and deliver new values, including elements destructive to the existing business model. Business model innovation is in itself a learning process of combining and recombining capabilities in new ways to deliver value (Kaplan 2012, p. 121). An adjacent innovation platform allows the expansion of the scope of the current business model or innovation or the connecting and combining of a different new technical market. In particular, it is easy for engineers to expand the current business model or innovation clearly from a technical perspective or to create a new innovation. "The adjacent possible is a kind of shadow future hovering on the edges of the present state of things, a map of all the ways in which the present can reinvent itself." Thus, ideas from engineers will be high-value works of bricolage.

13.2.2.2 Value Proposition for the Engineer Open Innovation Business Model Circle

Value Proposition Template: Engineer Open Innovation Business Model Circle

1. Apply a new technology from the point of view of an engineer to the existing market.
   - In particular, it may be easy for engineers to access open innovation-friendly industries or sectors. As open innovation is an open connection between technology and the market, it is easy for engineers with new ideas in technology to develop new business models.

2. When new technology is applied to the existing market, describe the values additionally created or the values of a solution to a problem not solved by the existing method.
   - The values newly created by applying the ideas of engineers are described. They are recommended following business model innovation principles.

3. Among the values additionally listed or solved, describe the combinations. First, combine similar values. Second, combine values which are differentiated from existing ones. Third, combine values with very special meanings.
   - During this stage in which the new value proposition is finally materialized, make use of the access of the adjacent value or an adjacent business model.
   - Note that the adjacent value shows the value of the genuine engineer business model only when the existing value proposition is aggressively destructed.

Table 13.6 Enablers of barriers of radical innovation

Enablers or barriers	Contents
Forces that generate and nurture breakthrough innovation	Partnering with outside companies to create new ventures Management commitment to support ideas outside the current strategy Availability of resources to support breakthrough ideas
Barriers to breakthrough innovation	Incentives focused on avoiding risk Breakthrough ideas are difficult to implement in manufacturing and distribution Perceived competition with existing businesses

Source: Davila et al. (2012, p. 53)

13.2.3 How Should Firms Combine New Technology and a New Market?

13.2.3.1 Key Factors

Enablers and Blockers of Radical Innovation

By accelerating radical innovation, the innovation and development of business models can be rapidly promoted. To promote radical innovation, it is important to establish a new start-up by partnering with an external company, actively supporting and implementing external ideas rather than current strategies, and securing resources to promote innovative ideas actively, such as those listed in Table 13.6. However, elements such as risk aversion, difficulty in implementing innovative ideas, and conflicts between new innovative business models and the existing business model make the introduction of a radical innovation, that is, a new and creative business model, difficult. Thus, when building the technological system, it is necessary to consider the factors that promote the introduction of radical innovation and creative business models in a system as well as those that block the introduction of creative business models when the system is designed.

Similarities Between Technology Levers and Business Model Levers

As shown in Table 13.7, the value proposition, value chain, and target of business model levers correspond to product service, process technology, and the enabling of the technology of technology levers, respectively. That is, the three levers target the same object but accessed from different aspects. In the six elements, a small change in any creates incremental innovation. However, when the two levers on both sides of business model and the technology levers, that is, value proposition and the product and service, value chain, and process technology, along with the target customer and enabling technology, change significantly together, radical innovation then occurs.

Thus, when engineers design technological systems, the promotion of important changes of technology levers and their business model levers should be considered to build more creative business models.

Table 13.7 Similarities between business model and technology levers and types of innovation

	Levers		
	Business model levers		
	Value proposition	Value chain	Target customer
	Product and service	Process technology	Enabling technology
Types of innovation	Technology levers		
Incremental innovation	Small changes in one or more of the six levers		
Semi-radical	Significant change in business model levers or technology levers		
Radical	Significant change in business model levers and technology levers, especially in small areas		

Source: Davila et al. (2012, p. 41) modified

Balance Between Creativity and Value Capturing

Considering that start-ups have creativity and that mature companies focus on capturing value, it can be predicted that engineers who place emphasis on new technologies or ideas tend to focus on creativity. In particular, creativity and value capturing should be balanced when engineers pursue open innovation between technology and the market with creative business model development. This balance must be concretely applied during the technological system design stage.

13.2.3.2 Technological System for the Engineer Open Innovation Business Model Circle

Technological System Template: Engineer Open Innovation Business Model Circle

1. Materialize a concrete figure of customers newly emerge in the value proposition stage through a new technology introduced by an engineer in a new market.
   - During this process, promote partnering to allow a new business model to accelerate breakthrough innovations and avoid passive actions of risk aversion.

2. Materialize a new technology proposed by an engineer based on new markets suggested by him/her.
   - Realize more creative innovation through significant changes of technology levers and their corresponding business model levers.

(continued)

3. Suggest a concrete combination method of a new materialized market and technology.
   - To balance creativity and value capture, a concrete combination method of technology and the market needs to be designed.

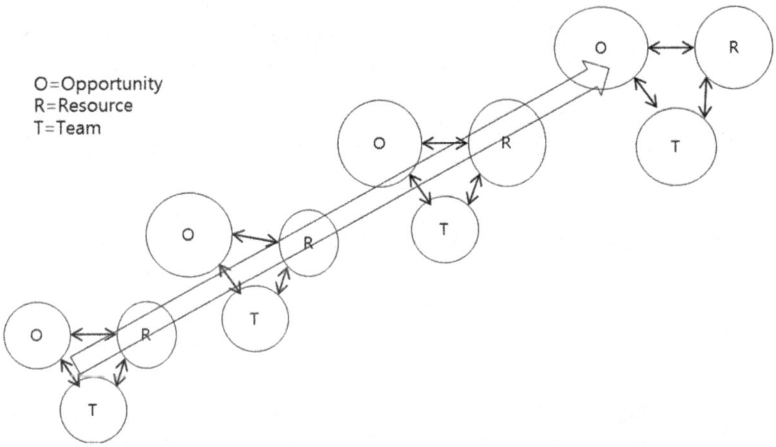

Fig. 13.3 The process of dynamic balance among opportunities, resources, and teams (Source: Timmons and Spinelli (1994, pp. 92–93) modified)

13.2.4 Why Should Firms Observe This Combination?

13.2.4.1 Key Factors

Dynamics Among Opportunities, Resources, and Teams

As shown in Fig. 13.3, new start-ups find ways newly to maintain balance in the dynamic process that began with the acquisition of strong opportunities after the balance of opportunities, resources, and teams. Based on the market segment acquired by combining new technology and the market, which represents a new opportunity in itself, through the dynamic process of securing resources, such as financial resources, assets, and people, and expanding new teams with full capabilities, a new business model is introduced to secure the balance in another dimension, which differs from the initial case.

In particular, from the perspective of engineers, a creative combination of technology and the market creates a new business model when the massive acquirement of opportunities creates a new balance among opportunities, resources, and teams through a dynamic process. Engineers should keep in mind that capturing or

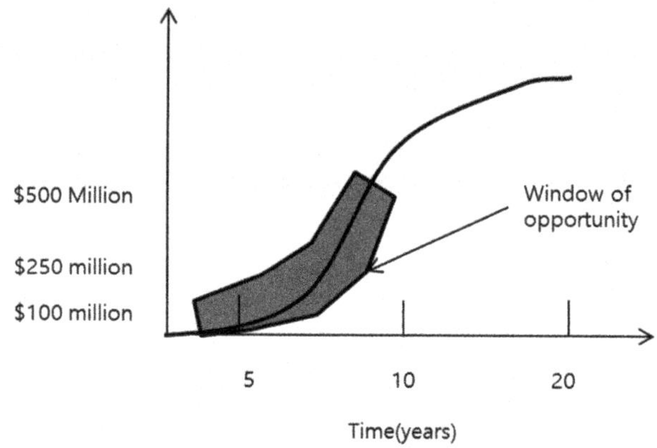

Fig. 13.4 Window of opportunity in technology-based industries (Source: Timmons and Spinelli (1994, p. 126) modified)

acquiring opportunities does not secure a new level of balance. Hence, a new opportunity should lead to investments on a new scale of resources, the securing of such an investment, and finally the expansion of new levels of teams to ensure that the new business model is sustainable.

Capturing the Window of Opportunity

As shown in Fig. 13.4, for a new technology-based business, the period of 5–10 years of activity and a business size of USD 100–500 million are sufficient for a window of opportunity. Under these conditions, if engineers with new technical ideas enter a market, they will do so with in good timing because a technical start-up can survive with its new business model in the market with a minimum size or larger in terms of market size and because the industry is not yet mature, thus allowing new start-ups to escape from being under the control of larger companies through economies of scale. Thus, when engineers form a new business model based on their technology, if they design their revenues and costs, the target industry should set the direction that corresponds to the window of opportunity.

13.2.4.2 Costs and Revenues for the Engineer Open Innovation Business Model Circle

Cost and Revenue Template: Engineer Open Innovation Business Model Circle

1. Clearly suggest the reason, structure, size, and description of cost cutting.
   - As the start of an engineer business model involves the application of a new technology to an existing market, typically, the element of cost reduction is included.

(continued)

- In particular, clarify the aspect of creating cost reductions in the medium and long term through new opportunities.
- Set the direction to allow the window of opportunity to be USD 1–5 million and for a new business to be a full-fledged business within 10 years after its introduction to the market.

2. Clearly suggest the reason, structure, size, and description of revenue creation caused by the application of a new technology.
   - Do not stop with merely suggesting and applying an opportunity. Rather, also secure minimum conditions of sustainability through investments in new resources.
3. Concretely consider additional designs of cost and revenue additionally to create synergistic effects through the creative combination of cost reduction and revenue creation.
   - In particular, have a creative synergy effect for cost reduction and new revenue creation by creating new levels of team combinations in addition to securing resources with new opportunities.

13.2.5 When and Where Does One Meet New or Additional Customers?

13.2.5.1 Key Factors

Quadrants Between a Channel and a Product

Engineers should clarify whether a product is physical or bit/virtual and whether a channel through which to interact with consumers is online (Web) or off-line (physical), as shown in Fig. 13.5. Hence, the location in the quadrants between channel and product to which one's product is placed should be clarified. The cases shown in Fig. 13.6 are mere examples; these cases have not been confirmed thus far, as they remain highly flexible. Based on Fig. 13.5, engineers should confirm the characteristics of their products and the channels to meet their consumers, and they need to think about which customer relations and channels should be set.

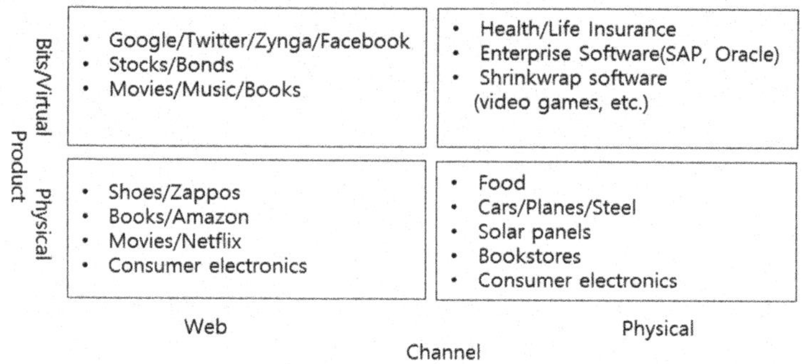

Fig. 13.5 The relationships between channels and products (Source: Blank and Dorf (2012, p. 17) modified)

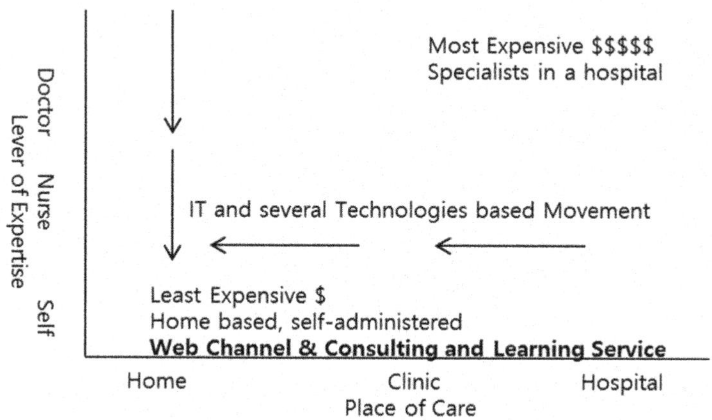

Fig. 13.6 Changes in health-care customer channels (Source: Christiansen et al. (2008, p. 13) modified)

The Changes of Channels in Health Care

As shown in Fig. 13.6, the development of ICT significantly changes channels and customer relations in the area of health-care services. Conventionally, customers receive medical services by seeing a doctor at a hospital or clinic. At present, health-care services are largely given through an app or over the Internet for consulting or information acquisition. That is, the introduction of ICT in the health-care field revolutionized customer relations and channels.

Changes in the Channels of Education

The activation of ICT leads to significant changes in customer channels and relations in the education industry, as shown in Fig. 13.7. Many online and mobile education platforms that supplement and substitute conventional educational institutions, such as colleges, provide many advanced education services, and they have been skyrocketing in number around the world. As engineers pay attention to the development of business models based on the introduction of new technologies such as ICT, new customer channels and relations should be fully considered and reviewed.

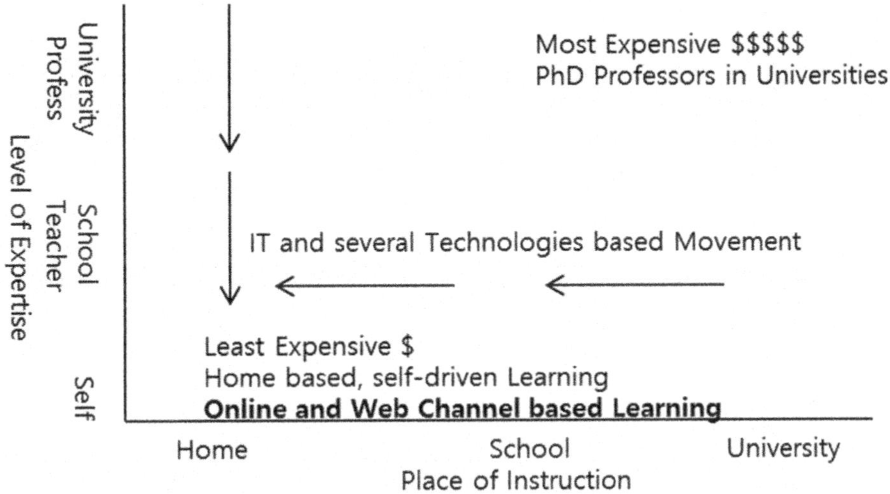

Fig. 13.7 Changes in education customer channels (Source: Chutani et al. (2010, p. 64))

13.2.5.2 Channels and Customer Relations for the Engineer Open Innovation Business Model Circle

Channels and Customer Relation Template: Engineer Open Innovation Business Model Circle

1. Design a technically creative customer relationship.
   - Fully review and apply the change of the customer relationship caused by the technology introduced by engineers in advance.
   - In addition, materialize the creative customer relationship to be operated in a new market.

2. Design a technically creative customer channel.
   - Fully review and apply the change of the customer channel caused by the technology introduced by engineers in advance.
   - In addition, materialize the creative customer channel to be operated in a new market.

3. Find new customer relations and a customer channel necessary creatively to operate the combination between new technology and the market.

Research Question

1. Select a case firm from newspaper, and analyze the business model of it from engineer open innovation-based business model developing circle.
2. Develop your own business model from engineer open innovation-based business model developing circle with the hypothesis that you are an engineer.

References

Acha V (2008) Open by design: the role of design in open innovation. Acad Manag Proc 2008(1):1–6
Auletta K (2010) Googled: the end of the world as we know it. Random House, New York
Blank S, Dorf B (2012) The startup owner's manual. K&S Ranch, Pescadero
Brin S, Page L (1998) Anatomy of a large-scale hypertextual Web search engine. 7th international world wide web conference
Chesbrough HW, Teece DJ (2002) Organizing for innovation: when is virtual virtuous? Harvard Business School Press, Boston
Christiansen C, Grossman JH, Hwang J (2008) The innovator's prescription. McGraw-Hill, New York
Chutani S, Aalami JR, Badshah A (2010) Technology at the margins: how it meets the needs of emerging markets, vol 22. Wiley, Hoboken
Davila T, Epstein M, Shelton R (2012) Making innovation work: how to manage it, measure it, and profit from it. FT press, Upper Saddle River
Gassmann O (2006) Opening up the innovation process: towards an agenda. R&D Manag 36(3):223–228
Herstad S (2007) Global open innovation: market learning, related variety and the global-local interplay in Norwegian industry. Oslo, NIFU STEP report
Jarvis J (2011) What would Google do?: reverse-engineering the fastest growing company in the history of the world. Harper Business, New York
Kaplan S (2012) The business model innovation factory: how to stay relevant when the world is changing. Wiley, Hoboken
Morris C (2014) Tesla motors: how Elon musk and company made electric cars cool, and sparked the next tech revolution. Smashwords Edition, Los Gatos
Page L (2001) Method for node ranking in a linked database: Google Patents.
Patra SK, Krishna VV (2015) Globalization of R&D and open innovation: linkages of foreign R&D centers in India. J Open Innov Technol Market Complex 1(1):1–24
Schmidt E, Rosenberg J (2014) How Google works. Hachette UK, London
Tietze F (2012) Technology market transactions: auctions, intermediaries and innovation. Edward Elgar Publishing, Northampton
Timmons JA, Spinelli S (1994) New venture creation: entrepreneurship for the 21st century, vol 4. Irwin, Burr Ridge
竹内一正. (2013). 未来を変える天才経営者イーロン・マスクの野望: 株式会社朝日新聞出版

Social Entrepreneur Open Innovation-Based Business Model Developing Circle

14

Abstract

This chapter discusses cases and the development processes and methods for social entrepreneur open innovation (OI)-based business models (BMs). To start, two business model cases for social entrepreneurs are introduced.

Afterward, the five steps in the social entrepreneur business model circle are explained with key factors and templates.

Keywords

Social entrepreneur • Burro battery • Microfinance in Pakistan • Social entrepreneur business model circle

14.1 Cases

14.1.1 Burro Battery

After receiving stock options and retiring from Microsoft as a veteran, Whit Alexander cofounded "Cranium" with his colleague, Richard Tait, in 1997, and cocreated the Cranium board game. He then sold the company to Hasbro, a well-known American toy company, for USD 75 million in 2007. After the sale, he established a battery company in Ghana in 2008 called "Burro." Alexander's brother Max, a journalist, and Tait also joined the foundation process. In addition, American companies took part in Burro's establishment as investors.

Burro has the mind-set of a social company that pursues social value but with a clear goal of creating such value based on market value itself. "Do More," the vision of the cofounders, carries the responsibility of providing more work opportunities for people in Africa as the company, as well as its profits, grows. They design and

J.J. Yun, *Business Model Design Compass*, Management for Professionals,
DOI 10.1007/978-981-10-4128-0_14

Table 14.1 Market and social values of Burro

Career of founder	After working for Microsoft, he received stock options and retired from the firm He cofounded Cranium, a game company, and operated it for 10 years, after which he sold the company to Hasbro, a toy company, for USD 75 million
Market value	He loans long-lasting batteries to people in Africa for half the usual price as a means of profit creation Burro continuously innovates with regard to its batteries, sales system, and distribution system to lead the qualitative and quantitative development of the battery loan business and to increase profits
Social value	By giving jobs to locals in Africa, Burro provides them with work experience In addition, Burro provides products to people to allow them to work with the right tools and earn more
Relationship between the two values	Burro creates and expands social value, being led by market value and based on the implementation of social value

implement their business to create a mutual virtual circle that carries their goal to sell and loan many Burro batteries to develop as a company with the ability to provide numerous opportunities for Ghanaians to participate with regard to job and value creation. To be specific, the mutual implementation of both market and social values is Burro's ultimate goal. By creating a significant amount of profit, Burro offers work opportunities for Ghanaians which, ultimately, shines a light on their condition. The cofounders wanted to create a brand that focuses on products and services that improve productivity and that low-income families can afford.

Burro upholds three values in its everyday operations: respect, innovate, and empower (Alexander 2012, p. 267). First, Burro respects its agents in Ghana, and its agents respect the company's customers. Second, Burro continuously innovates with regard to battery performance and its sales system, loan types, and management system by always seeking the best methods. The company shares its good ideas to make them a source of new innovation, keeps track of customers' requirements and expectations, and continuously innovates to improve its batteries, sales system, and corporate operations based on the information gathered. Third, Burro encourages its customers to work harder and provides them tools to enable them to do so to achieve its main goal, i.e., to help people "Do More." By providing quality batteries for half the normal price, the company allows Ghanaians to be more productive. The creation of "hybrid value chains" will allow the for-profit and citizen sectors to work together with the goal of reshaping the economy and realizing social change with a long-term impact (Drayton and Budinich 2010). Social entrepreneurs do not emerge out of a vacuum but exist in vast multitudes (P. Sen 2007). Under the absolute poverty and lack of infrastructure in Africa, the founders of Burro continue to make efforts to achieve the creation of both market and social value.

Led by market value but driven by social value, Burro successfully creates both, as shown in Table 14.1. A social entrepreneur accumulates the experience and capacity of market value creation through a start-up foundation and mergers and acquisitions (M&A) and continues to create social value based on the accumulated market value.

14.1.2 Microfinance in Bangladesh

Microfinance is an expanding program in which a small loan system lends money, generally ranging from USD 100 to USD 300, to the poor so that they can start their own businesses. In fact, Grameen Bank (GB), Building Resources Across Communities (BRAC), Proshika: A Centre for Human Development (Proshika), the Association for Social Advancement (ASA), and others are engaged in the microfinance business in Bangladesh (Karim 2011, p. vii). According to Bateman and Chang (2009), although "the widespread assumption that simply 'reaching the poor' with microcredit will automatically establish a sustainable economic and social development trajectory animated by the poor themselves… the microfinance model may well generate some positive short run outcomes for a lucky few of the 'entrepreneurial poor,' the longer run aggregate development outcome very much remains moot."

Microfinance in Bangladesh to some extent follows the concept of microcredit as a socioeconomic program, with the only difference being that it is based on entrepreneurship rather than public assistance. With this, it is estimated that close to 7.6 million households in 1997, 100 million households in 2006, and 175 million households in 2015 received the benefits of this system. As Yunus admits, a social enterprise can handle most social and economic problems, such as those in medical and financial services, information technology, education and training, marketing, and renewable energy; the problem, according to Yunus, is to innovate with a business model (BM) efficiently to achieve the desired social result (Yunus 2007, p. 233). However, GB neither provides business model consulting services nor sets business models as important standards of lending money. Instead, the bank lets money borrowers recite the "16 decisions of Grameen Bank," which focus on capitalistic and individualistic standards, diligence, self-control, hygiene, saving, and other values, and concentrates on "(1) built-in financial safeguards; (2) the community as fiscal enforcers; and (3) the instrumentalization of shame as a loan recovery technology" (Karim 2011, p. 73).

To overcome poverty, methods of securing measures to be enacted by the poor themselves or those to eliminate political and economic inequality are suggested through democratic discussions and debate which become an alternative to microfinance (Karim 2011, p. 204; Royce 2015, p. 247). To do this concretely, however, people in poverty need to devise an innovative business model based on a creative link and a combination of technology and the market or society based on sound alternatives (Yun 2015). Unfortunately, the reality is that microfinance in Bangladesh is inexperienced with regard to the components of business models, including know-how, knowledge, experience, or consulting as well.

The microfinance business in Bangladesh highlights social value itself without any links to market value; thus, social value is not implemented in the end; with

Table 14.2 Microfinance in Bangladesh

Overview	A small loan system that lends money to the poor without security and allows them to realize a new business and repay the money with their business profits
Market value	The microfinance system in Bangladesh has a loan rate of around 10%. This interest is lower than other rates in the country but significantly high under the current global financial situation. Thus, this high rate becomes a strong incentive to encourage global banks to enter the microfinance field of the nation
Social value	The original purpose of the microfinance system is to provide a starting point to the poor to overcome their economic difficulties through their efforts. However, the program is operated not by concretely combining with business models but strongly redeeming a loan based on human relations. Thus, the social purpose of this system is not achieved, and this program simply becomes a different form of loan sharking
Relation between the two values	While the side effects of the market value of a borrower and the desired social value are combined, microfinance skyrockets. However, its social value is not fully achieved, and multiple side effects occur

this, non-implementation, unexpected side effects arise. That is, this system is discounted as another type of loan sharking, as shown in Table 14.2. With this, it is difficult for social value not directly combined with market value to be implemented by a social firm (Khandker 2003).

14.2 Social Entrepreneur Open Innovation-Based Business Model Developing Circle

As shown in Fig. 14.1, if a social entrepreneur creatively combines technology and the market by approaching the circle according to the principle of the five Ws and one H, he or she can create a new business model (Yun 2015; Yun et al. 2015). This business model will, above all else, be developed based on the connection to and combination of technology with the social market. Martin and Osberg (2007) provide three components of social entrepreneurship, as follows:

1. Identifying a stable but inherently unjust equilibrium that causes the exclusion, marginalization, or suffering of a segment of humanity that lacks the financial means or political clout to achieve any transformative benefit on its own
2. Identifying an opportunity in this unjust equilibrium, developing a social value proposition, and bringing to bear inspiration, creativity, direct action, courage, and fortitude, thereby challenging the stable state's hegemony
3. Forging a new, stable equilibrium that releases trapped potential or alleviates the suffering of the targeted group and through imitation and the creation of a stable ecosystem around the new equilibrium, ensuring a better future for the targeted group and even society at large

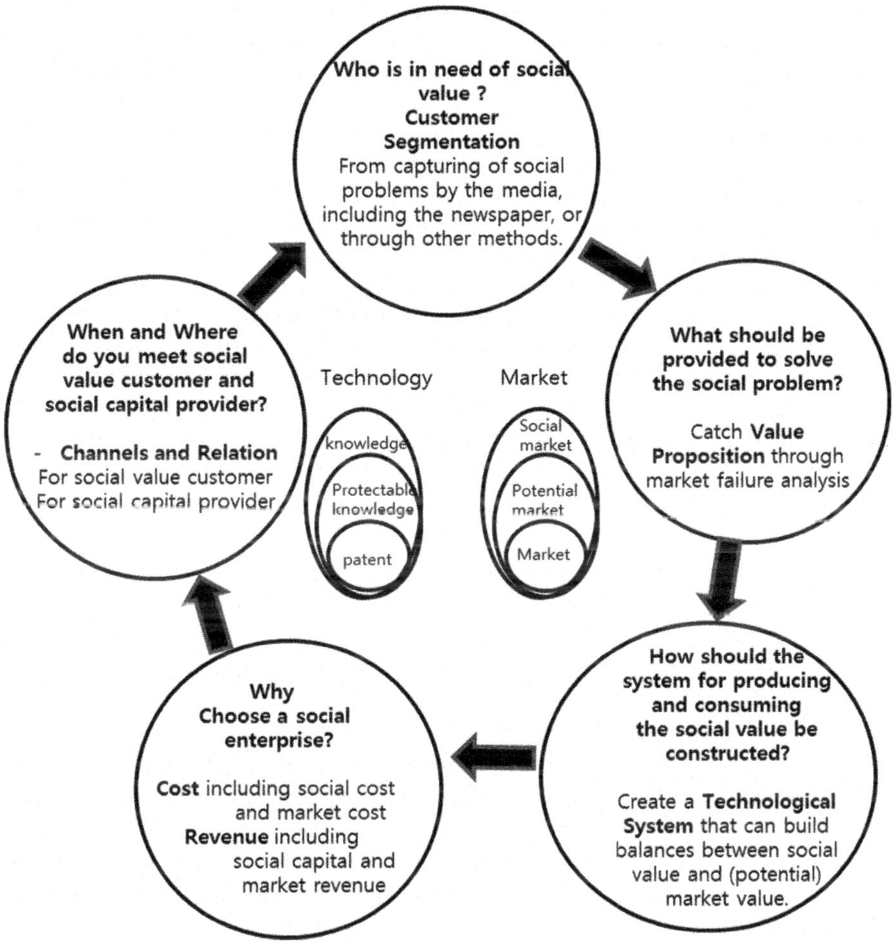

Fig. 14.1 Social entrepreneur open innovation-based business model developing circle

Thus, an open business model developed from the perspective of a social entrepreneur differs from the existing three business models because the open business model pursues the realization of both market and social values. Chesbrough and Di Minin (2014) define open social innovation as "the application of either inbound or outbound open innovation (OI) strategies, along with innovations in the associated Business Model of the organization, to social challenges." That is, a business model created by applying open innovation to social innovation is a social entrepreneur-based open innovation business model.

If social entrepreneurs scale up their operations through the social entrepreneur open innovation-based business model developing circle, social innovation will generate a sufficient impact.

Table 14.3 Business models for nonprofits or social enterprises

Any organization that wants to be relevant, to deliver value at scale, and to sustain itself must clearly articulate and evolve its business model
Although core revenue to support operations of most nonprofits in the social sector comes primarily from grants and donations, trying to establish sustainable sources of service revenue is viewed as an alternative or as gravy on top of the core
Any nonprofit wholly dependent on grants to support its business model is at risk
Another problem with nonprofit business models that are dependent solely on grants to sustain themselves is the constraints that come with nearly every grant
A business model innovation for the public sector, nonprofits, or social enterprises will mitigate the fear and risk of change by demonstrating what new models may look like when tested on a smaller scale in a less threatening connected adjacency

Source: Kaplan (2012, pp. 167–178)

14.2.1 Who Is in Need of Social Value?

14.2.1.1 Key Points

Nonprofit Enterprises Have Business Models as Well

The information presented in Table 14.3 shows precisely how social enterprises cannot survive without business models. Thus, even social enterprises based on grants or donations essentially need to develop creative business models.

Technology Follows Ideas or Social Business Models

"If technology does become more responsive to public needs rather than state needs and business needs, the influence will be partly mediated through ideas" (Mulgan 2015, p. 169). To respond to the needs of the public or society rather than the urgent needs of the market, an idea that combines a specific technology with social demand is necessary. That is, there is a need for social and economic business models that are created through the combination of technology and the social market. In particular, technology follows ideas or social business models, unlike in the cases of existing markets such as health care, the relational economy, the green economy, and civilization, which measure social value.

14.2.1.2 Customer Segmentation for the Social Entrepreneur Open Innovation Business Model Circle

Customer Segmentation Template: Social Entrepreneur Open Innovation Business Model Circle

1. Capture social problems through social issues covered by newspapers or TV news.
   - It is recommended to deliberate over the reasons and appropriateness of social enterprises' failures to survive without business models.

(continued)

2. Check if the problems need to be handled through social methods rather than the market.
   - First, propose a social business mode, and then identify which technology is needed to implement the business model.
3. Clearly and completely define the targets achieved from the solution to the social problem.

14.2.2 What Should Be Provided to Solve a Social Problem?

14.2.2.1 Key Factors

Social Entrepreneurship That Can Capture Social Value

As listed in Table 14.4, social entrepreneurs, with their type of entrepreneurship, aim to identify and solve social problems that have not been solved in the market. Through this process, they focus on increasing social benefits rather than individual benefits. However, social entrepreneurs do not oppose the market, instead focusing on solutions to social problems not solved by the market while taking market methods into consideration.

The Entitlement Approach to Social Problems

"The entitlement approach concentrates on each person's entitlements to commodity bundles, including food, and views starvation as resulting from a failure to be entitled to a bundle with enough food" (Sen 1981, p. 45). The creative combination of society and the technology that can overcome the existing entitlement through the entitlement approach to social problems can be a source of innovative and new value proposition. Amartya Sen, who was awarded the Nobel Prize in Economic Sciences, conducted research which involved an internal analysis that found that great famines around the world, such as the Great Bengal Famine of 1943, the famines in Ethiopia, the drought and famine in the Sahel, and the Bangladesh famine of 1974, took place due to entitlement rather than natural phenomena. However, by approaching many social problems faced by people in their everyday lives as well as famine from the perspective of entitlement, it is possible to have gain creative ideas to handle social problems and form a new social value propositions.

Table 14.4 Social entrepreneurship

Social entrepreneurship addresses social problems, or it responds to needs that are unmet by private markets or governments
Social entrepreneurship is motivated primarily by social benefits
Social entrepreneurship generally works with—not against—market forces

Source: Brooks (2009, pp. 4–5)

14.2.2.2 Value Propositions for the Social Entrepreneur Open Innovation Business Model Circle

Value Proposition Template: Social Entrepreneur Open Innovation Business Model Circle

1. Describe why values are not created, distributed, and consumed through the market business between suppliers and consumers.
   - Starting with the awareness and value of social entrepreneurship, describe why problems are not solved through market methods, including concrete contents, in many different dimensions.
2. Clarify the values to be offered as a social value rather than market value. That is, describe the value that is socially significant despite the fact that it does not directly provide any rewards for the value offering.
   - Clearly provide more creative and diverse social values through the entitlement approach.
3. Clarify the entire social market value to offer this value and the value by companies continuously or by combining the market value to be offered in combination with social value as the basis of greater social value creation and provision.

14.2.3 How Should the System for Producing and Consuming Social Value Be Constructed?

14.2.3.1 Key Factors

The World Is Remarkably Unequal

The industrial revolution started in England in the eighteenth century and sparked great economic growth, which became the driving force that allowed many people to overcome their physical poverty; however, it also became the cause of the Great Divergence (Deaton 2013, p. 4). The Great Divergence is the separation of countries in Northern Europe, Western Europe, and North America from other countries; this period also created an enormous gap between Western countries and the rest of the world. As capital income was significantly greater than labor income in the late twentieth century and in the twenty-first century, the inequality between people in

Table 14.5 The development process of social business models (Source: Brooks (2009, p. 7)) modified

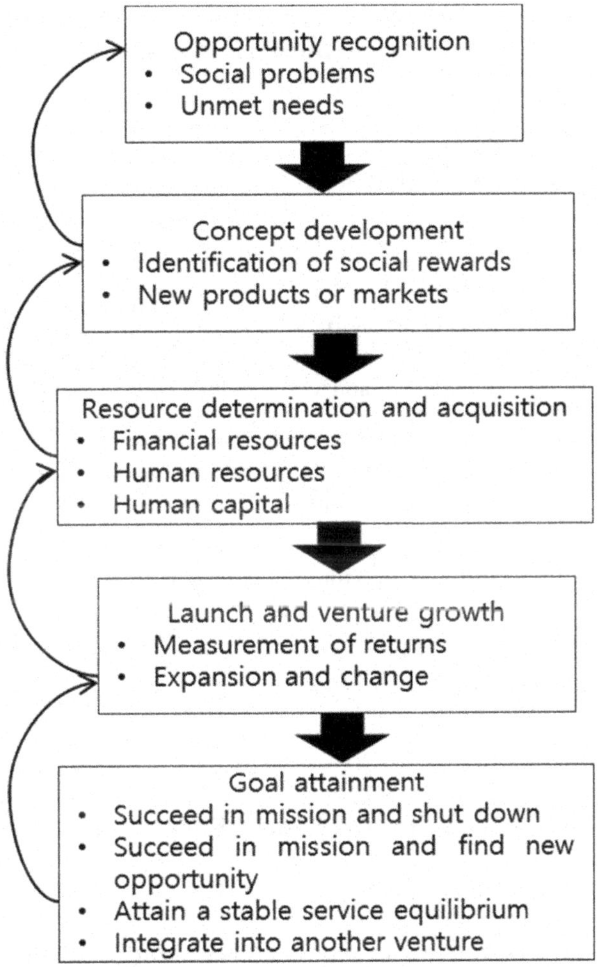

addition to that among nations has accelerated (Piketty 2014, p. 237). To reduce this acceleration, various social approaches should initially be considered, starting with national or social intervention in the commons and the knowledge commons (Frischmann et al. 2014; Ostrom 2015).

The Process of Social Business Model Development

The development of a social business model starts with the awareness of social problems or unmet needs, conceptualizes social rewards, and goes through the steps laid out in Table 14.5. However, each step does not pull a person in one direction, and, at times, the process returns to the previous step. In fact, if necessary, a specific step continues for a longer period. The structural characteristics that meet social requirements distinguish consumers from social needs. And it displays the satisfaction of social consumption which is led by a nonmarket process.

14.2.3.2 Technological System for the Social Entrepreneur Open Innovation Business Model Circle

Technological System Template: Social Entrepreneur Open Innovation Business Model Circle

1. Design the system to provide social value in a simple yet concrete manner.
   - Pay attention to the fact that the inequality generated from the development of capitalistic societies leads national, regional, and individual inequality, and contemplate how to handle the side effects of such inequality.
2. Design the system to provide market value, which is the basis of offering social value, in a simple yet concrete manner.
   - Suggest the application of the fact that social business model is created in stages and through feedback.
3. Integrate the entire system to realize the virtuous circle and a sustainable balance between social value and market value production distribution systems.

14.2.4 Why Choose a Social Enterprise?

14.2.4.1 Key Factors

Social Return on Investment Beyond the Social Value Proposition

The components of the social return on investment (SROI) include a financial return on investment (enterprise value), costs and savings from serving one's social mission (social purpose value), and the economic and socioeconomic value of enterprise (blended value). That is, the revenue of social companies satisfies both enterprise and social purpose values.

The Zero-marginal-cost Society

From market capitalism to collaborative commons, an era with an entirely new cost concept is emerging. That is, from the new combination of technology and the market by an entrepreneur to the creative destruction through technology innovations based on the R&D investments of large enterprises to the marginal cost, zero is set as the product price. This zero-marginal-cost society brings about technological unemployment and also the collaborative commons by social capital (Rifkin 2014, p. 24), not as the tragedy of the commons but as the comedy of the commons. The trend is exemplified as sharing open-source S/W through copyleft, the Energy Commons Society, the Communication Commons Society, and the renaissance of unions.

Finding Capitalism's Generative Ideas

What makes the sharing economy or common economy different is how it rediscovers collective intelligence and cooperation and empathy, which are ignored in the traditional capitalistic market. In a new economic era, the value of a collaborative economy is maximized in various ways through a perfect community rather than a perfect market. Entrepreneurship beyond business shows new characteristics that creatively respond to social values or needs (Mulgan 2015, p. 47).

14.2.4.2 Cost and Revenue for the Social Entrepreneur Open Innovation Business Model Circle

Cost and Revenue Template: Social Entrepreneur Open Innovation Business Model Circle

1. Clearly define the market cost and revenue.
   - Distinguish the enterprise value and social purpose value of a company, and clarify the market cost concept.
2. Clearly define social cost and revenue.
   - Consider the characteristics of the zero-marginal-cost society, and derive the social cost and revenue of a business model.
3. Design the balance among market cost, revenue cost, and revenue and cost.
   - Short-term social revenue and cost > market revenue and cost
   - Long-term market revenue and cost > social revenue and cost
   - (The continuous growth of social revenue over the long term is based on the market revenue.)
   - Fully considering the practical characteristics which arise due to the emergence of the sharing economy and the common economy, pursue a long-term balance of the type in which market revenue leads social revenue.

14.2.5 When and Where Does One Meet Social Value Customers or Social Capital Providers?

14.2.5.1 Key Factors

Let Us Become the Adaptable Giver with a Generous Tit-for-tat Strategy

According to A. Sen (1973, p. 1), "that a perceived sense of inequity is a common ingredient of rebellion in societies is clear enough, but it is also important to recognize that the perception of inequity, and indeed the content of that elusive concept, depend substantially on possibilities of actual rebellion." The process of decreasing inequity is to create the driving force of the creation of the start-up open innovation economy through dynamics, finally leading to a new closed innovation economy with mass production (Yun 2015). To do this, one must initially build the collaboration and dynamics of giving and taking credit when meeting customers with social value or contributors during the value creation process. Second, through the power of powerless communication, it becomes possible to meet customers with social value. Third, one must maintain motivation by matching concern for self-interest and that for others' interests. Fourth and last, through the strategy of generous tit-for-tat beyond tit-for-tat, one will meet customers with social value and participants who produce social value (Grant 2013, p. 195).

Socioeconomic Development Sharing and Collaboration

As shown in the following table, various companies that focus on sharing form a sharing platform on which to meet social customers and market customers. In addition, companies related to the collaboration economy create platforms to collaborate with social customers concretely and meet market customers. That is, the sharing and collaboration economy firms allow customers with social value and the analogous market value to participate in creating and consuming necessary values through various types of platforms. It is these types of systems which have skyrocketed around the world (Table 14.6).

Table 14.6 Examples of firms that exemplify the sharing and collaboration economy

Sharing or collaboration target	Example of company
Car sharing	PhillyCarShare, City CarShare, HOURCAR
House sharing	Airbnb, HomeAway, Couchsurfing
Toy sharing	BabyPlays, Rent That Toy!, Sparkbox Toys
Tie sharing	Tie Society
Women clothing sharing	Rent the Runway
Babies and children's clothing sharing	Thredup
Unused product sharing	Yerdle
Unused real estate sharing	Shared Earth
Illness care information collection collaboration	PatientsLikeMe, CureTogether
Social capital crowdfunding	Zopa, Lending Club, Prosper, Kickstarter, Indiegogo, EarlyShares, Crowdfunder, Fundable, Crowdcube
Review site	Yelp, Angie's List, Citysearch, TripAdvisor, Travelocity, Judy's Book

Source: Rifkin (2014, pp. 251–438), Tapscott and Williams (2008) summarized

14.2.5.2 Channels and Customer Relationship for the Social Entrepreneur Open Innovation Business Model Circle

Channels and Customer Relationship Template: Social Entrepreneur Open Innovation Business Model Circle

1. When and where do we meet social value customers?
   - Which channels can be accessed to meet customers with social value?
   - How can we form a relationship with customers with social value and maintain it?
   - Create an approach from the perspective of an adaptive giver, that is, a generous tit-for-tat relationship.
2. When and where do we meet market value customers?
   - What is the connection channel with customers after the marketization of social value?
   - How can we form a relationship with customers after the marketization of social value?
   - Concretely develop the channel and customer relationship by reviewing the platforms of various companies of the sharing and collaboration economy.
3. When and where do we meet social capital providers?
   - How can we build channels with social capital providers?
   - How can we form a relationship with social capital providers?

Research Question

1. Select one example firm which produces social value, and analyze its business model with social entrepreneurship-based open innovation and business model.
2. Build up your own social entrepreneurship-based open innovation having business model to conquer social problems which you are interested in.

References

Alexander M (2012) Bright lights, no city : an African adventure on bad roads with a brother and a very weird business plan. Hyperion, New York

Bateman M, Chang HJ (2009) The microfinance illusion. University of Juraj Dobrila Pula and University of Cambridge.

Brooks AC (2009) Social entrepreneurship: a modern approach to social value creation. Pearson Prentice Hall, Harlow

Chesbrough H, Di Minin A (2014) Open social innovation. New Frontiers Open Innov:169–188

Deaton A (2013) The great escape: health, wealth, and the origins of inequality. Princeton University Press, Princeton

Drayton B, Budinich V (2010) A new alliance for global change. Harv Bus Rev 88(9):56–64

Frischmann BM, Madison MJ, Strandburg KJ (2014) Governing knowledge commons. Oxford University Press on Demand, Oxford

Grant A (2013) Give and take: a revolutionary approach to success. Hachette UK, London

Kaplan S (2012) The business model innovation factory: how to stay relevant when the world is changing. Wiley, Hoboken

Karim L (2011) Microfinance and its discontents: women in debt in Bangladesh. University of Minnesota Press, Minneapolis

Khandker SR (2003) Micro-finance and poverty: evidence using panel data from Bangladesh. World Bank Policy Research Working Paper (2945).

Martin RL, Osberg S (2007) Social entrepreneurship: the case for definition. Stanf Soc Innov Rev 5(2):28–39

Mulgan G (2015) The locust and the bee: predators and creators in capitalism's future. Princeton University Press, Princeton

Ostrom E (2015) Governing the commons. Cambridge University Press, Cambridge

Piketty T (2014) Capital in the twenty-first century. Harvard University Press, Cambridge

Rifkin J (2014) The zero marginal cost society: the internet of things, the collaborative commons, and the eclipse of capitalism. Macmillan, New York

Royce E (2015) Poverty and power: the problem of structural inequality. Rowman & Littlefield, Lanham

Sen A (1973) On economic inequality. Oxford University Press, Oxford

Sen A (1981) Poverty and famines: an essay on entitlement and deprivation. Oxford University Press, Oxford

Sen P (2007) Ashoka's big idea: transforming the world through social entrepreneurship. Futures 39(5):534–553

Tapscott D, Williams AD (2008) Wikinomics: how mass collaboration changes everything. Penguin, New York

Yun JJ (2015) How do we conquer the growth limits of capitalism? Schumpeterian dynamics of open innovation. J Open Innov Technol Market Complex 1(1):1–20

Yun JJ, Jeong E, Yang J (2015) Open innovation of knowledge cities. J Open Innov Technol Market Complex 1(1):1–20

Yunus M (2007) Creating a world without poverty: social business and the future of capitalism. PublicAffairs, New York

Index

J.J. Yun, *Business Model Design Compass*, Management for Professionals,
DOI 10.1007/978-981-10-4128-0

Lightning Source UK Ltd.
Milton Keynes UK
UKOW01n0607220817
307696UK00007B/60/P

9 789811 041266